Body Composition and Aging

Interdisciplinary Topics in Gerontology

Vol. 37

Series Editors

P.R. Hof New York, N.Y.
C.V. Mobbs New York, N.Y.

Body Composition and Aging

Volume Editors

C.V. Mobbs New York, N.Y.

P.R. Hof New York, N.Y.

23 figures and 5 tables, 2010

KARGER

Basel · Freiburg · Paris · London · New York · Bangalore · Bangkok · Shanghai · Singapore · Tokyo · Sydney

Charles V. Mobbs, PhD
Department of Neuroscience
Mount Sinai School of Medicine
New York, N.Y., USA

Patrick R. Hof, MD
Department of Neuroscience
Mount Sinai School of Medicine
New York, N.Y., USA

Library of Congress Cataloging-in-Publication Data

Body composition and aging / volume editors, C.V. Mobbs, P.R. Hof.
p. ; cm. -- (Interdisciplinary topics in gerontology, ISSN 0074-1132
; v. 37)
Includes bibliographical references and index.
ISBN 978-3-8055-9521-6 (hard cover : alk. paper) -- ISBN 978-3-8055-9522-3
(e-ISBN)
1. Human body--Composition. 2. Adipose tissues. 3. Aging--Physiological
aspects. I. Mobbs, Charles V. II. Hof, Patrick R. III. Series:
Interdisciplinary topics in gerontology ; v. 37. 0074-1132
[DNLM: 1. Body Composition--physiology. 2. Aged. 3. Aging--physiology.
4. Obesity--physiopathology. 5. Sarcopenia--physiopathology. W1 IN679
v. 37 2010 / QU 100 B6673 2010]
QP86.B577 2010
612--dc22

2010022259

Bibliographic Indices. This publication is listed in bibliographic services, including Current Contents® and Index Medicus.

www.karger.com
Printed in Switzerland on acid-free and non-aging paper (ISO 9706) by Reinhardt Druck, Basel
ISSN 0074–1132
ISBN 978–3–8055–9521–6
e-ISBN 978–3–8055–9522–3

Contents

Preface

The present volume is a natural successor to a previous volume (vol. 35) in this series, 'Mechanisms of dietary restriction in aging and disease'. In that volume, the role of body composition in mediating age-related impairments constituted a constant context, but was only addressed obliquely. Certainly, the obesity epidemic has made the role of body composition human health ever more salient, but there have been surprisingly few books surveying this subject, and none, to our knowledge, focusing on animal models examining the causes and consequences of age-related changes in body composition. Yet, as clearly described in the first chapter, increased adiposity with age may play a key role in determining lifespan. On the other hand, as described in the second chapter, the role of adiposity and mortality during aging is much more complex than it would appear, and in fact increased adiposity may play a paradoxically protective role to increase pathology yet decrease mortality in the elderly. Similarly, the third and fourth chapters describe the converse dangers of decreased body mass index secondary to reduced appetite during aging, which may be at least as deleterious as obesity. The next two chapters describe that the two salient changes in body composition during human aging, increased adiposity and decreased muscle mass, also characterize species as diverse as rats and *Caenorhabditis elegans*, suggesting that these changes may constitute an almost universal feature of aging in animals. This conclusion is supported by the following three chapters, which describe the robustness of sarcopenia with age and address several potential mechanisms addressing this apparently universal aspect of aging in animals. The next chapter addresses the key question of the efficacy of exercise in ameliorating age-related impairments, usefully contrasting these effects to those of dietary restriction. Finally, the causes and consequences of the third major age-related change in body composition, bone loss, are described. We are indebted to the outstanding investigators who have contributed their time and efforts in writing these reviews, and hope that our readers benefit as much from reading the articles as we did from editing them.

Charles V. Mobbs, Patrick R. Hof
New York, N.Y., USA

Mobbs CV, Hof PR (eds): Body Composition and Aging.
Interdiscipl Top Gerontol. Basel, Karger, 2010, vol 37, pp 1–19

Contribution of Adipose Tissue to Health Span and Longevity

Derek M. Huffman[a,c] · Nir Barzilai[a–c]

Departments of [a]Medicine, [b]Genetics and [c]Institute for Aging Research, Albert Einstein College of Medicine, Bronx, N.Y., USA

Abstract

Adipose tissue accounts for approximately 20% (lean) to >50% (in extreme obesity) of body mass and is biologically active through its secretion of numerous peptides and release and storage of nutrients such as free fatty acids. Studies in rodents and humans have revealed that body fat distribution, including visceral fat (VF), subcutaneous (SC) fat and ectopic fat are critical for determining the risk posed by obesity. Specific depletion or expansion of the VF depot using genetic or surgical strategies in animal models has proven to have direct effects on metabolic characteristics and disease risk. In humans, there is compelling evidence that abdominal obesity most strongly predicts mortality risk, while in rats, surgical removal of VF improves mean and maximum life span. There is also growing evidence that fat deposition in ectopic depots such as skeletal muscle and liver can cause lipotoxicity and impair insulin action. Conversely, expansion of SC adipose tissue may confer protection from metabolic derangements by serving as a 'metabolic sink' to limit both systemic lipids and the accrual of visceral and ectopic fat. Treatments targeting the prevention of fat accrual in these harmful depots should be considered as a primary target for improving human health span and longevity.

Obesity and Disease

The prevalence of overweight (body mass index; BMI >25) and obesity (BMI >30) now affects nearly two-thirds of Americans and has reached epidemic proportions in most of the developed world. The fundamental cause of obesity is a long-term imbalance in energy intake and expenditure (i.e. positive energy balance) leading to increased body mass including the accumulation of subcutaneous (SC) and visceral fat (VF). Although general obesity is an important risk factor for many diseases, several human studies have demonstrated that VF accrual, which is a hallmark of aging in humans [1], as a strong predictor of many health complications. In this review, we will discuss (1) data in humans demonstrating the importance of body fat distribution in the determination of disease risk, (2) studies which demonstrate a causal relationship between VF rather than SC fat accumulation to disease risk, and (3) corollary and causative evidence that VF unfavorably impacts mortality risk and longevity.

Visceral Fat Accrual as a Determinant of Disease Risk in Human
Obesity increases the risk for several comorbidities including type 2 diabetes (T2DM), stroke, cardiovascular disease (CVD), and metabolic syndrome [2]. In a report from the National Health and Nutrition Examination Survey III (NHANES III), by comparison with normal weight men, the odds of having metabolic syndrome increase in a dose-response manner from overweight (5.2), moderately obese (25.2), to severely obese (67.7) [3]. More recently, the risks associated with obesity have been extended to many site-specific cancers, including prostate [4]. Relevant to this review, studies have shown that lifestyle interventions which combat general and abdominal obesity [5] can lessen the risk of developing and dying from these diseases which could significantly impact both human health span and longevity.

Over the past two decades, a growing number of studies have also assessed the health risk attributed to abdominal obesity either by using waist circumference or waist-to-hip ratio for example or by utilizing state-of-the-art imaging tools. Remarkably, the majority of these studies have concluded that central obesity is a stronger risk factor for disease and mortality than BMI, total body fat or other fat depots. Indeed, several cardiometabolic complications are increased by visceral obesity, including insulin resistance, T2DM [6], CVD [7], heart failure [8] and stroke [9]. Interestingly, the hazards associated with abdominal obesity have even been extended to central nervous system disorders such as cognitive decline, [10], Alzheimer's disease [11] depression [12] and dementia [10].

Studies have also found that an increasing waist circumference is predictive of worsening mobility and disability [13] as well as frailty [14] in elderly men and women. Likewise, several recent studies have linked visceral adiposity with greater risk for developing colorectal cancer [15] while a limited amount of data have found associations with cancers of the liver [16] and prostate [17]. It has also been observed that SC and VF have differential and, in some cases, opposing effects on health. For example, it was recently shown that increasing amounts of SC fat were associated with better bone structure and strength while VF was negatively linked with these outcomes [18]. It is not entirely clear how visceral obesity is causally related to these diseases, but its link to insulin resistance and secretion of pro-inflammatory markers, as will be discussed in the following section, have been proposed [19].

Potential Mechanisms Linking Visceral Fat to Disease Risk in Humans
The mechanism(s) linking VF with the metabolic syndrome has been proposed to involve its anatomical location, leading to a 'portal' effect of greater free fatty acids (FFA) and glycerol release with proximal access to the liver [20]. This hypothesis had

remained attractive for many years since obesity is associated with higher FFA levels and excess circulating FFA are known to promote insulin resistance in skeletal muscle and liver [21, 22]. However, careful examination of FFA fluxes in obese humans did not strongly support this model and have raised some doubts as to the importance of FFA in linking VF to insulin resistance [23]. Specifically, while an increasing amount of VF was found to augment the contribution of hepatic FFA delivery from VF, it was generally observed that VF was only a minor contributor to systemic FFA release (20% of portal vein FFA and 14% of total FFA release) [23]. Overall, this study did not lend strong support to the hypothesis that portal release of FFA from VF explains its link to hepatic insulin resistance.

More recently, evidence has shown that adipose tissue is an active endocrine organ, capable of secreting many cytokines, often referred to as adipokines, that can promote a pro-inflammatory state and interfere with insulin action [24]. For example, human studies have found that VF predicts C-reactive protein (CRP) and interleukin-6 (IL-6) levels in young and middle-aged men and that adjusting for VF eliminates the increase observed in these pro-inflammatory markers commonly observed with aging [25]. Likewise, Fontana et al. [26] directly measured adipokine arteriovenous level differences across fat and concluded that VF is an important site of IL-6 secretion. Furthermore, studies from our group [27, 28] and others [29] have shown that SC and VF are biologically distinct, with VF demonstrating far greater pro-inflammatory characteristics.

Ectopic Fat Deposition and Disease

In addition to visceral adiposity, fat accumulation in ectopic depots, including liver, muscle, pancreas and heart have also proven highly relevant to disease risk [30, 31]. It is not entirely clear what leads to fat accumulation in these organs rather than adipose tissue, but it likely involves the net effect of excess FFA flux, increased lipogenesis or a decreased ability to locally oxidize lipids, leading to excess fat deposition. Ultimately, excess lipid accrual in these metabolically relevant tissues can result in lipotoxicity, although it is unclear if this is the direct result of TG or potentially more toxic intermediates such as diacylglycerol and ceramides [31, 32], which among other things may interfere with tissue insulin signaling. Obese, type 2 diabetic subjects have been observed to have increased intramyocellular lipid (IMCL) content and the amount of IMCL content is predictive of the degree of insulin resistance, even after adjustment of visceral and total adiposity [33]. Furthermore, mice predisposed to accrue IMCL in either muscle or liver by tissue-specific overexpression of lipoprotein lipase demonstrate impaired insulin sensitivity in muscle and liver, respectively [34].

Fat accumulation in liver (i.e. fatty liver) is another well-characterized site of ectopic fat accrual and can lead to fatty liver disease and even more severe complications including, steatohepatitis, necrosis, and hepatocellular carcinoma [35, 36]. Interestingly, the accumulation of VF has been shown to predict intrahepatic triglyceride (IHTG) content [37] and an increase in IHTG is also associated with many of the same metabolic abnormalities linked to visceral adiposity, including hyperlipidemia, insulin resistance and T2DM [38]. Since visceral obesity is correlated with IHTG content, a recent study [39] postulated that fat accrual in liver rather than VF, is a better marker of metabolic disease in nondiabetic humans. In order to discern between these two sites of fat storage, obese subjects were grouped by high or low IHTG content, but matched for VF. Alternatively, subjects were also categorized by VF volume (high or low) but matched on IHTG content, and insulin sensitivity was assessed by hyperinsulinemic-euglycemic clamps and very-low-density lipoprotein-triglyceride (VLDL-TG) secretion rate was assessed with tracers. Interestingly, they observed that when subjects were matched for VF, insulin action at the level of the liver, adipose tissue and skeletal muscle was significantly impaired while the VLDL-TG secretion rate was concomitantly increased in subjects with the greatest IHTG content [39]. In contrast, when individuals were matched for IHTG, no differences in insulin sensitivity or VLDL-TG secretion were observed between subjects with high versus low VAT volume. The study conclusion was that hepatic fat accumulation in humans trumps visceral adiposity as a predictor of obesity-related metabolic dysfunction. However, others have shown that VF accrual is predictive of the extent of hepatic inflammation, fibrosis and steatohepatitis in humans [40], such that the inflammatory potential of VF may still be an important catalyst for derangements occurring in liver.

In summary, there is abundant evidence which links obesity, and specifically visceral obesity, as an independent predictor of systemic inflammatory markers, hyperlipidemia and numerous diseases including cardiometabolic risk, cognitive decline and cancer in humans. There are also data in rodents and humans linking fat accumulation in ectopic depots, including liver and skeletal muscle, with metabolic abnormalities such as insulin resistance. Although not discussed in detail here, ectopic fat accumulation in heart and pancreas has also been linked to metabolic complications, while a recent study has even implicated a role for perivascular fat in obesity as a source of local inflammation and poor vascular tone [41]. Therefore, while VF remains the single strongest predictor of cardiometabolic risk and other complications, ectopic fat accrual may also be highly relevant. Because many health complications are associated with obesity, more studies are necessary to better understand both the individual contribution and biology of central and ectopic fat accrual to metabolic derangements in humans.

Animal Models of Visceral Obesity

Human studies have clearly demonstrated that obesity, and specifically abdominal obesity is strongly associated with disease risk [42]. In order to better understand the pathophysiology of obesity in mammals, several strategies have been developed and implemented including, high-fat feeding in rodents [43] and dogs [22], seasonal models of obesity [44], transgenic mice [45] and spontaneous mutants such as *ob/ob* and *db/db* mice [46]. A common feature of obese animal models is a marked increase in VF and hyperinsulinemia, but similar to human studies, the distinct contribution of body fat distribution in these models cannot be directly elucidated. However, clever experimental strategies in animal models including transgenic [47] and surgical approaches [19, 48] have been conducted in order to demonstrate causality between VF and disease. The next section will briefly highlight key studies which have employed either fat transplantation or removal in animal models in an effort to demonstrate a direct link between VF and disease.

Fat Transplantation Models

Ohman et al. [49] determined whether transplanting epididymal fat pads from C57BL/6J mice could affect vascular disease in atherosclerosis-prone apolipoprotein E-deficient (ApoE–/–) mice. In plasma from ApoE–/– mice receiving fat transplants, they displayed elevated levels of leptin, resistin, and monocyte chemoattractant protein-1 (MCP-1) compared with sham-operated mice. Furthermore, mice transplanted with VF developed significantly more atherosclerosis than controls. In contrast, SC fat transplants did not accelerate atherosclerosis despite a similar degree of inflammation, suggesting that VF rather than SC fat-related inflammation accelerates atherosclerosis development in this model. Likewise, another study designed to determine whether differences in SC and VF are intrinsic or due to anatomic location transplanted either SC or VF pads into recipient mice [50]. They found no effect on the metabolic phenotype when VF was transplanted into recipient mice. However, they did observe that when SC fat was transplanted into the visceral cavity, recipient mice lost body weight and body fat and had improved insulin sensitivity, suggesting that SC fat is intrinsically different from VF. Furthermore, this study is in agreement with the growing contention that SC fat is protective from metabolic abnormalities as will be discussed in more detail later.

Surgical Removal of Visceral Fat

Our group and others have selectively removed perinephric and epididymal fat pads from rats and mice and studied its impact on insulin action (refer to table 1 for a summary of the metabolic effects of VF removal in animal models). In an initial study [48] from our laboratory, young moderately obese Sprague-Dawley (SD) rats were randomized either to surgical removal of VF (VF–) or to sham operation (SO). Since the VF that was removed accounted for only ~10% of total fat mass, there was no difference between groups for body weight or total fat mass. Nevertheless, VF– rats demonstrated significantly reduced insulin levels, and during a glucose clamp, the rates of insulin infusion required to maintain plasma glucose levels and hepatic glucose production were reduced by 50%. More recently, we have shown that VF accretion is an important determinant of hepatic insulin resistance during pregnancy, which can be largely prevented by surgical removal of VF prior to mating [51], providing further evidence that VF directly impacts hepatic insulin action.

Table 1. Effect of visceral fat removal on metabolic parameters in animal models

Study	Species/strain	Glucose/insulin	Cytokines	Lipids	Insulin action
Barzilai et al. [48]	male SD rats	↔ glucose ↓ insulin 49%	↓ TNF-α 75% and ↓ leptin 60% expression in SC fat	↔ FFA ↔ glycerol	↓ insulin infusion rate 53% ↓ IGFBP-1 68%
Kim et al. [111]	female SD rats (MSG treated)	↔ glucose ↔ insulin	NA	↓ FFA 66% ↓ TG 18% ↔ cholesterol	NA
Gabriely et al. [19]	male SD rats	↔ glucose ↓ insulin 39%	NA	↔ FFA	↑ glucose infusion rate 80% ↑ glucose uptake 30% ↓ glucose production 50%
Gabriely et al. [19]	male ZDF rats	↓ insulin 50%	NA	↓ FFA 44% during clamp	↑ glucose infusion rate ↔ glucose uptake ↓ glucose production 20%
Borst et al. [52]	male SD rats	↔ insulin	↓ IL-6 53%, ↔ TNF-α ↓ adiponectin 30% ↓ resistin 26% ↔ leptin	↔ FFA	↓ GTT AUC 15% ↑ insulin-stimulated glucose transport in muscle 24–31%
Pitombo et al. [53]	male Swiss mice	↓ glucose 35% ↓ insulin 45%	↓ TNF-α 40% ↓ IL-6 60% ↑ adiponectin 40%	NA	↑ insulin action 45%
Einstein et al. [51]	female pregnant SD rats	↔ glucose ↔ insulin	↔ IL-6 ↔ adiponectin	↓ hepatic TG 60% ↓ FFA 38%	↑ GIR 32%, ↓ HGP 42% ↔ glucose uptake
Lottati et al. [56]	male Mongrel dogs (Omentectomy)	↓ glucose 3% ↔ insulin	NA	↔ FFA	↔ HGP ↑ glucose uptake 68%
Shi et al. [54]	male C57BL6/J mice (Epididymal)	↔ glucose* ↓ insulin 7%	NA	NA	↓ GTT AUC 22% ↔ ITT
Shi et al. [54]	male FVBN mice				
(epididymal)	↓ glucose ↑ insulin 60%	NA	NA	↓ GTT AUC 51% ↔ ITT	

Table 1. Continued

Study	Species/strain	Glucose/insulin	Cytokines	Lipids	Insulin action
Shi et al. [54]	female C57BL6/J mice (parametrial)	↔ glucose ↔ insulin	NA	NA	↓ GTT AUC 40% ↔ ITT
Shi et al. [54]	female FVBN mice (parametrial)	↔ glucose ↔ insulin	NA	NA	↓ GTT AUC 23% ↔ ITT

* Glucose levels significantly decreased postsurgery but were not different from sham controls.

Borst et al. [52] studied the impact of VF removal on insulin action and skeletal muscle glucose transport in male SD rats. VF removal in rats tended to improve glucose tolerance (15% reduction in glucose AUC) and significantly lowered some pro-inflammatory adipokines in serum. Most striking, VF– animals displayed increased insulin-stimulated glucose transport in excised soleus and digitorum longus muscle as compared to sham-operated controls. Overall, these studies provide verification that VF is a potent modulator of both hepatic and peripheral insulin action.

Our group [19] and others [53] have also demonstrated protection from T2DM in rodents with VF removal. In our study, 2-month-old Zucker diabetic fatty (ZDF) rats were assigned to receive either a sham operation (ZDSO), or surgical removal of VF (ZDVF–) [19]. Despite no differences in plasma glucose or FFA levels, ZDVF rats had a marked reduction in fasting insulin levels by ~50%. In addition, during a glucose clamp, ZDVF rats demonstrated a greater glucose infusion rate and insulin suppression of glucose production than controls. Furthermore, when rats were monitored longitudinally, the development of diabetes, as determined by fasting glucose, was delayed in ZDVF- rats but not in sham-operated rats.

Using a mouse model of diet-induced obesity and T2DM, Pitombo et al. [53] assessed the impact of VF removal on glucose metabolism, insulin signaling and serum adipokine levels. Control mice became diabetic and hyperinsulinemic, but VF removal partially restored metabolic parameters. In addition, VF removal completely attenuated the impairment of insulin signaling observed in skeletal muscle from control animals and normalized serum adipokines levels. Other studies have also shown a similar differential response to SC versus VF removal and that these responses are influenced by both the sex of the animal and diet [54, 55]. More recently, it was shown that greater omentectomy in dogs improved peripheral insulin sensitivity, a finding that was attributed to an albeit small (7%) reduction in VF [56]. In humans, evidence that VF removal can favorably affect metabolic parameters is limited, though one pilot study found that the addition of omentectomy to gastric banding resulting in lower fasting glucose and insulin [57]. Taken together, these studies demonstrate the ability of VF removal to improve insulin action and prevent T2DM in animal models.

Can Subcutaneous Fat Be Beneficial?

Surrogate measures of obesity including increased body weight, BMI and total body fat have long been known to be predictive of greater cardiometabolic risk and other health complications [58]. Despite the fact that over two-thirds of adult Americans are considered overweight or obese, not all of these individuals have characteristics of the metabolic syndrome [59]. In contrast to VF accretion, most but not all [60] studies in humans and rodents have found that SC fat is often at worst, mildly associated with risk while most have either found no link or even a protective effect of SC fat on metabolic parameters [61]. This has led to the growing hypothesis that SC fat may be beneficial by serving as a 'metabolic sink' to limit both systemic lipids and the accrual of fat in more harmful regions, such as visceral and ectopic fat depots.

Surgical Removal of Subcutaneous Fat in Humans and Rodents

Early studies on the effect of liposuction in humans, which targets the removal of large quantities of SC fat in the abdominal area, have produced conflicting results on metabolic outcomes. A study by Klein et al. [62] attempted to definitively address this controversy by performing liposuction in humans as a tool to treat obesity-related metabolic disorders. In a study consisting of 15 obese patients (8 nondiabetic controls and 7 T2DM patients) with similar BMI, several metabolic parameters were assessed before and 10–12 weeks after having ~10.5 kg of SC abdominal fat removed. Utilizing hyperinsulinemic-euglycemic clamps, they found that liposuction did not significantly alter insulin action in muscle, liver, or adipose tissue, nor was there any effect on plasma levels of CRP, IL-6, tumor necrosis factor-α (TNF-α), adiponectin, glucose, insulin, blood lipids and blood pressure. A more recent study from the same group also found no improvement in CVD risk factors, such as HDL, LDL and TG with liposuction [63]. Although the initial study from Klein and colleagues is now often cited as authoritative evidence against a role for SC fat in metabolic complications in humans, some studies continue to report improvements in cardiometabolic risk factors with liposuction [64, 65]. Why these studies have reached different conclusions regarding the effect of this procedure remains an issue of debate.

We have assessed the effect of SC fat removal in rats in order to understand which fat depot(s) plays a 'causative' role in the peripheral and hepatic insulin resistance of aging. As described previously, VF extraction in old rats was able to restore peripheral and hepatic insulin action to the levels observed in young rats [19]. Interestingly, surgically removing an equivalent amount of SC fat (SC–) had no effect on peripheral or hepatic insulin sensitivity. In addition, while postabsorptive plasma glucose and FFA levels were similar in all groups, plasma insulin levels were ~50% higher in the SC– and sham-operated rats compared with VF rats. Thus, surgically removing substantial amounts of SC fat in rodents fails to improve markers of metabolic health.

Expansion of Subcutaneous Adipose Tissue Improves Metabolic Parameters

Kim et al. [66] hypothesized that a major link between obesity and insulin resistance is due to a limited ability to expand SC adipose tissue leading to fat deposition in the viscera and ectopic fat deposition in skeletal muscle and liver. In order to test this hypothesis, they overexpressed adiponectin in *ob/ob* mice which lack leptin. They observed that increasing adiponectin levels dramatically improved the metabolic profile in *ob/ob* mice including normalized glucose and insulin levels and lower TG. They also displayed reduced macrophage infiltration in adipose tissue and systemic inflammation. Most striking, the transgenic mice were morbidly obese, weighing nearly 100 g by 20 weeks of age as compared to *ob/ob* littermates that weighed nearly 40 g less. Furthermore, transgenic mice had much greater fat mass including SC fat mass, but less relative VF (VF% of total body weight) than *ob/ob* mice and less TG in muscle and liver. These data illustrate an interesting paradox which is in line with most studies showing no metabolic benefit to SC fat removal. This finding is also in agreement with the study of Tran et al. [50] described earlier demonstrating metabolic benefits from transplanting SC fat pads into donor mice. Collectively, these studies suggest that expansion of the SC fat depot is not harmful but instead appears to be beneficial for insulin action by preventing ectopic and VF deposition.

Expansion of SC fat as a therapeutic target for T2DM treatment has also been proposed in humans. As an example, compounds that activate the transcription factor, peroxisome proliferator-activated receptor-γ (PPAR-γ), such as thiazolidinediones (TZDs), have been commonly used to treat insulin resistance and T2DM in humans [67]. It is now known that these compounds can promote depletion of circulating lipids [68] and ectopic fat stores [69] in humans and promote differentiation and expansion of preadipocytes in vitro [70]. More recently, a study examining the effect of pioglitazone on body fat distribution, insulin sensitivity and fat cell size in obese humans found that 12 weeks of treatment increased the proportion of SC fat and reduced VF stores [71]. This is also in line with prior studies demonstrating that incubation of preadipocytes from human SC and omental fat with TZDs increased lipid accumulation in the SC, but not omental adipocytes [72].

Increased Subcutaneous Fat Mass in Long-Lived Mice

Life span extension has been demonstrated in animal models with diminished insulin/IGF-1 signaling and/or reduced function of the somatotropic axis, including the Ames *(Prop1^{df})* and Snell dwarf mouse *(Pit1^{dw})*, or mice with a disruption in the GH receptor-binding protein gene (GHR–/–) [73]. These mouse models are also commonly observed to be substantially smaller than wild-type and to have improved insulin sensitivity. Interestingly, despite their small body size, these mice have been shown to have similar levels of absolute fat mass and hence greater percentage fat than wild-type mice. However, the accumulation of fat mass in these mice preferentially occurs in the SC depots in both males and females, with concomitantly low levels of VF [74].

Obesity and Mortality Risk in Humans

In light of the overwhelming evidence linking obesity to disease risk, it should come as no surprise that obesity has also been shown to increase the risk of disease-specific and all-cause mortality [75–78] and reduce human life expectancy [79]. For example, a landmark study by Calle et al. [80] in 2003 reported that obesity was associated with increased risk of death from all cancers and many site-specific cancers in men and women including, liver, pancreas, stomach, esophagus, colon, breast, prostate, gallbladder, non-hodgkin's lymphoma, and multiple myeloma. Furthermore, it was estimated that 90,000 deaths due to cancer each year in the US could be prevented by maintaining a normal weight. Likewise, obesity has been linked to a reduced life expectancy [79] and to accelerated aging as demonstrated by obese women having shorter telomeres (~240 bp) than age-matched lean controls [81].

Abdominal Obesity and Mortality Risk

Since abdominal obesity, as assessed by waist circumference or waist-to-hip ratio for example, in large population studies has emerged as a stronger predictor of disease risk than BMI, studies have begun assessing the mortality risk posed by abdominal obesity [82–84]. While results from these studies have varied, inconsistencies have been attributed to different methods of obtaining and expressing anthropometry surrogate data, as well as the sex, ethnicity and age of the population. There have been several notable recent studies which have demonstrated that abdominal obesity measures most strongly predict mortality risk. For instance, Wannamethe et al. [85] found that a particularly high waist circumference (>102 cm), waist-to-hip ratio (top quartile), and a composite of waist circumference and sarcopenia were the strongest predictor of mortality in men. Similarly, a large cohort study in Europe (EPIC) reported that general (BMI) and abdominal adiposity (waist circumference; waist-to-hip ratio) were both strong predictors of mortality risk but that the importance of abdominal obesity was most striking among persons with a low BMI [86]. Likewise, the incidence of overall and cardiovascular-related death was greatest in those with a low BMI coupled with a high waist circumference [87].

General Obesity, Abdominal Obesity and Mortality: Effect of Age

When considering the impact of obesity on mortality risk and life expectancy, it is important to also consider other variables such as the sex, race and age of the population. For example, in a study by Fontaine et al. [79], it was shown that the earlier age at which individuals became obese, the greater the toll on life expectancy, and this was generally true regardless of sex or race. For example, Caucasian men with a BMI

>45 in their 20s, could expect a reduction in life expectancy of 13 years while those who reached this level in the seventh decade of life would only expect to lose 2 years off their life expectancy. Other studies have also shown that the health risk associated with a high BMI declines with advancing age, while inversely, a low BMI in the elderly has actually been shown to be predictive of a greater mortality risk [88].

In regards to abdominal obesity, there have only been a limited number of studies performed which have assessed the toll of abdominal obesity in relatively younger versus older populations on life expectancy. While the impact of abdominal obesity in middle-aged individuals has proven detrimental for disease-specific and all-cause mortality risk in most studies, the consequence of abdominally obesity on mortality risk in the elderly has been mixed. For instance, it was shown in a study utilizing the NHANES data set, that out of several body composition measures, abdominal obesity most strongly predicted mortality risk, but this observation was only detected in adults younger than 65 years [89]. Likewise, a study by Reis et al. [88] found that most surrogate measures of increased abdominal obesity in middle-aged (30–64 years) men and women predicted mortality risk, but not in those >64 years of age. In fact, a greater BMI in both sexes and WC in men were actually associated with a lower risk of mortality. Likewise, in a 12-year follow-up study of nearly 5,800 men and 6,400 women >75 years of age, Price et al. [90] found that a higher BMI or WC was associated with lower mortality risk in men and women while a modest increase in mortality risk was observed with an increasing waist-to-hip ratio. On the other hand, the Health Professionals Study [91] reported that a greater waist circumference or waist-to-hip ratio, in men >65 years was predictive of coronary heart disease risk or cardiovascular disease-related mortality.

Taken together, there is sufficient evidence to warrant abdominal obesity as an important risk factor not only for disease risk but also mortality risk in young and middle-aged populations. However, the relationship between abdominal obesity and mortality in the elderly is more ambiguous. There are several possibilities which may explain a lack of consistency in studies of this population and these have been described in detail elsewhere [90]. Briefly, surrogate measures of abdominal obesity may be confounded by age-related changes in body composition including the loss of both SC fat and lean mass. It is also not clear whether simple measures such as waist circumference or ratios are adequate or appropriate estimates of abdominal obesity in the elderly, and as these studies often rely on a single measurement, it is not known how long individuals have had abdominal obesity. As weight loss and fat loss are also a common observation in the elderly prior to death, the reliability of the measurement as representative of their peak late-life abdominal obesity should also be considered. Along those same lines, people with large quantities of VF may already have died prior to becoming elderly leading to a selection bias in the population. Finally, the possibility that abdominal obesity in the elderly is a risk factor for disease-specific deaths, such as CVD-related deaths, rather than all-cause deaths, is another possibility.

Role of Adiposity in Caloric Restriction-Mediated Longevity

Studies dating back to the early 1900s by Moreschi and Rous, and later by McCay and Tannenbaum [92, 93], respectively, were the first to demonstrate that a reduction in food intake was capable of increasing life span and inhibiting tumor formation in rats [94]. Nearly a century later, calorie restriction (CR) remains the only known behavioral intervention capable of delaying the onset of many age-related diseases and extending maximal longevity [95].

In the laboratory, CR is generally implemented by limiting food intake 20–40% of ad libitum-fed controls [96]. Thus, the beneficial effects of CR have historically been attributed to a reduction in food intake [97–99]. However, this view has since been challenged as CR is not only characterized by less food intake, but also by concurrent changes energy balance, body mass, and body composition [100, 101]. Because adipose tissue has been historically viewed as an inert storage depot for TG, this robust phenotypic change had been widely discounted as merely a byproduct of reducing energy intake [97, 98, 102]. It was not until 1960 that a reduction in fat stores was proposed as an important mediator of CR [103]. Since many believed CR worked by slowing metabolism or retarding growth and development, this hypothesis was never fully embraced at the time, although it was neither discounted until 20 years later.

Evidence against a Role for Body Fat in Determining Life span

In 1980 and 1984, respectively, two historic studies were published which essentially discredited the 'reduction of body fat hypothesis' for many years. The first by Bertrand et al. [97] reported that fat mass was not related to life span in ad libitum-fed rats, but was paradoxically associated with a greater life span in calorie-restricted rats. The later study by Harrison et al. [102] found that calorie-restricted *ob/ob* mice were longer lived than ad libitum-fed wild types, despite having nearly twice as much body fat. Indeed, these studies made a compelling case against a role for body fat in the determination of life span. However, the conclusions of these studies have since been criticized and debated [98, 101].

A Paradigm Shift: Adipose Tissue Is an Active Endocrine Organ

Since the discovery of leptin in 1994 [104], the view of adipose tissue has evolved from that as an inert storage depot to an active endocrine organ. As previously mentioned, obesity and specifically abdominal obesity is associated with a low-grade pro-inflammatory state [25]. Many of these cytokines also referred to as 'adipokines' including leptin, TNF-α, IL-6, PAI-1, heparin-binding epidermal growth factor (HB-EGF), and vascular endothelial growth factor (VEGF) among others, may play an important role in many disease pathologies by promoting angiogenesis, inflammation, cell proliferation, and insulin resistance [105, 106]. Considering this new perspective of fat, coupled with studies from our group linking age-related changes in body composition with insulin resistance, we hypothesized that the ability of CR to improve insulin action is primarily by reducing VF.

VF Removal in Rats Mimics the Benefits of Caloric Restriction on Insulin Action
Since CR limits the accumulation of VF and preserves hepatic insulin action with aging in mammals [107] and extends life span in a variety of species [108], we suspected that the ability of CR to prevent insulin resistance with aging was due to the attenuation of VF, rather than other fat depots. To directly test this hypothesis, we studied four groups of rats: VF–, SC– (equivalent SC fat removed), SO (sham-operated controls), and CR (CR+ sham operated) [19]. Postabsorptive plasma insulin levels were nearly 50% greater in the SC– and SO rats as compared to CR and VF– animals. During a glucose clamp, VF– rats had an 80% increase in the rate of glucose infusion, significantly greater glucose uptake, and 50% increase in the ability of insulin to suppress hepatic glucose production compared to SC– and SO rats. Most striking, the dramatic improvement in insulin action with VF, but not SC fat removal, closely resembled the effects of prolonged CR, suggesting that decreased VF could largely account for the beneficial effects of a reduction in food intake. Additionally, our earlier studies had shown a causal role for VF in age-related metabolic decline [101] leading us to revisit the body fat and longevity hypothesis by proposing that the ability of CR to improve longevity is by reducing VF.

VF Removal in Rats Improves Longevity
VF accretion is a common hallmark of aging, and we have demonstrated metabolic benefits to VF removal [107], and that decreased VF largely accounts for the improvement in insulin action with CR [19, 107]. Therefore, it seemed plausible that the beneficial effects of CR on longevity may be due to the attenuation of VF [101]. Quite surprisingly, very few rodent studies have found a role for body fat in the determination of life span. In the most widely cited report in support of a role for body fat by Bluher et al. [100] in 2003, fat-specific insulin receptor knockout (FIRKO) mice showed a modest reduction in body mass, but were markedly leaner than controls, demonstrating a 50% reduction in fat mass and lived nearly 20% longer than controls. Using a mathematical approach, Wang et al. [109] estimated that CR and body weight had independent effects on mortality rate in male Wistar rats, with body weight accounting for approximately 11% of the CR effect on mortality rate. However, this study may have underestimated the importance of body fat and specifically VF, since body composition was not measured.

In order to definitively demonstrate that VF modulates longevity, we prospectively studied life span in 3 groups of rats: ad libitum-fed (AL), 40% CR and VF– rats [110]. CR rats demonstrated the greatest survival and longevity among all experimental groups (fig. 1). Statistical analysis revealed a significant increase in both mean and maximum life span for VF– rats as compared to AL-fed animals and VF removal accounted for ~20% of the effect of CR on longevity. Furthermore, the hazard rate of death in the

Fig. 1. Mean and maximum life span of three groups of rats either fed ad libitum (AL), fed ad libitum with VF removed (VF–) CR rats had the greatest mean and maximum life span but VF removal only was sufficient to extend life span as compared to AL controls and accounted for nearly 20% of the CR effect on life span. Different letters denote a significant difference between groups, $p < 0.05$. The original survival curve can be found in Muzumdar et al. [110].

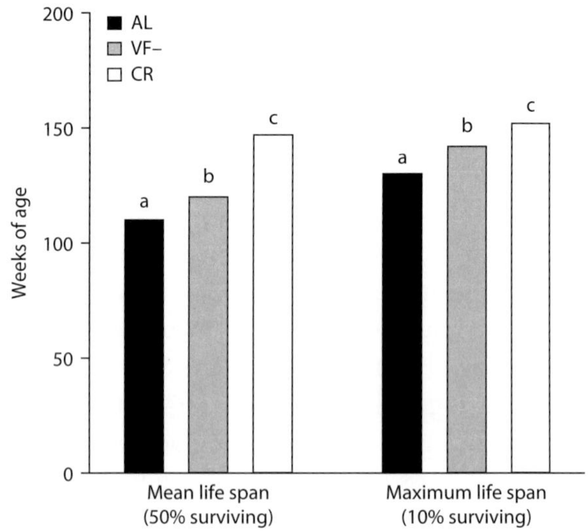

VF– group was 0.49 (51% reduction) and in the CR group was 0.13 (87% reduction) compared to the AL-fed group. This effect of VF *per se* was even more remarkable when considering that VF– rats had similar food intake, body weight and body fat as AL controls. In fact, VF– animals were 220 g heavier, had nearly 115 g more body fat and a percentage body fat nearly double that of CR rats. Furthermore, the only fat pads removed were the epididymal and perinephric fat pads, which do not drain into the portal vein in rats. Indeed, the mesenteric fat pad, which has been deemed the only 'true' VF pad in rodents because it drains into the portal vein, is difficult to safely remove in rodents. This raises the possibility in humans that VF depletion may be even more beneficial since VF depots in humans have direct portal access and hence a greater potential to harm the liver. In summary, these data provide the first causal evidence implicating VF depletion as an important underlying cause of improved life span with CR.

Conclusions

Studies in humans have implicated abdominal obesity as a major risk factor for insulin resistance, T2DM, CVD, stroke, metabolic syndrome and death. Studies in animal models have extended upon these observations to demonstrate a causal relationship between visceral adiposity and disease while surgical removal of VF in rats improves both mean and maximum life span. Fat deposition in ectopic depots such as pancreas, heart, skeletal muscle and liver can also lead to lipotoxicity and impair insulin action. Conversely, SC fat may be beneficial under certain circumstances by limiting levels of both circulating lipids and visceral and ectopic fat deposition. Taken together, these studies highlight the importance of accounting for body fat distribution when determining disease risk and the potential importance of preventing ectopic and VF accrual for improving human health-span and longevity.

Acknowledgements

This work was supported by grants from the National Institutes of Health (AG21654 and AG18381 to N.B.) and by the Core laboratories of the Albert Einstein Diabetes Research and Training Center (DK 20541). D.M.H. is supported by a postdoctoral T32 Training Grant (T32 AG23475) and a postdoctoral Fellowship Award from the Prevent Cancer Foundation.

References

1 Matsuzawa Y, Shimomura I, Nakamura T, Keno Y, Tokunaga K: Pathophysiology and pathogenesis of visceral fat obesity. Ann NY Acad Sci 1995;748:399–406.

2 Huffman DM, Barzilai, N: Role of visceral adipose tissue in aging. Biochim Biophys Acta 2009;1790: 1117–1123.

3 Park YW, Zhu S, Palaniappan L, Heshka S, Carnethon MR, Heymsfield SB: The metabolic syndrome: prevalence and associated risk factor findings in the US population from the Third National Health and Nutrition Examination Survey, 1988–1994. Arch Intern Med 2003;163:427–436.

4 Freedland SJ, Platz EA: Obesity and prostate cancer: making sense out of apparently conflicting data. Epidemiol Rev 2007.

5 Selwyn AP: Weight reduction and cardiovascular and metabolic disease prevention: clinical trial update. Am J Cardiol 2007;100:33P–37P.

6 Koh-Banerjee P, Wang Y, Hu FB, Spiegelman D, Willett WC, Rimm EB: Changes in body weight and body fat distribution as risk factors for clinical diabetes in US men. Am J Epidemiol 2004;159:1150–1159.

7 Pascot A, Lemieux S, Lemieux I, Prud'homme D, Tremblay A, Bouchard C, Nadeau A, Couillard C, Tchernof A, Bergeron J, Despres JP: Age-related increase in visceral adipose tissue and body fat and the metabolic risk profile of premenopausal women. Diabetes Care 1999;22:1471–1478.

8 Nicklas BJ, Cesari M, Penninx BW, Kritchevsky SB, Ding J, Newman A, Kitzman DW, Kanaya AM, Pahor M, Harris TB: Abdominal obesity is an independent risk factor for chronic heart failure in older people. J Am Geriatr Soc 2006;54:413–420.

9 Walker SP, Rimm EB, Ascherio A, Kawachi I, Stampfer MJ, Willett WC: Body size and fat distribution as predictors of stroke among US men. Am J Epidemiol 1996;144:1143–1150.

10 Whitmer RA, Gustafson DR, Barrett-Connor E, Haan MN, Gunderson EP, Yaffe K: Central obesity and increased risk of dementia more than three decades later. Neurology 2008;71:1057–1064.

11 Razay G, Vreugdenhil A, Wilcock G: Obesity, abdominal obesity and Alzheimer disease. Dement Geriatr Cogn Disord 2006;22:173–176.

12 Vogelzangs N, Kritchevsky SB, Beekman AT, Newman AB, Satterfield S, Simonsick EM, Yaffe K, Harris TB, Penninx BW: Depressive symptoms and change in abdominal obesity in older persons. Arch Gen Psychiatry 2008;65:1386–1393.

13 Guallar-Castillon P, Sagardui-Villamor J, Banegas JR, Graciani A, Fornes NS, Lopez Garcia E, Rodriguez-Artalejo F: Waist circumference as a predictor of disability among older adults. Obesity (Silver Spring) 2007;15:233–244.

14 Hubbard RE, Lang IA, Llewellyn DJ, Rockwood K: Frailty, body mass index, and abdominal obesity in older people. J Gerontol A Biol Sci Med Sci 2010; 65:377–381.

15 Yamaji T, Iwasaki M, Sasazuki S, Kurahashi N, Mutoh M, Yamamoto S, Suzuki M, Moriyama N, Wakabayashi K, Tsugane S: Visceral fat volume and the prevalence of colorectal adenoma. Am J Epidemiol 2009;170:1502–1511.

16 Ohki T, Tateishi R, Shiina S, Goto E, Sato T, Nakagawa H, Masuzaki R, Goto T, Hamamura K, Kanai F, Yoshida H, Kawabe T, Omata M: Visceral fat accumulation is an independent risk factor for hepatocellular carcinoma recurrence after curative treatment in patients with suspected NASH. Gut 2009;58:839–844.

17 von Hafe P, Pina F, Perez A, Tavares M, Barros H: Visceral fat accumulation as a risk factor for prostate cancer. Obes Res 2004;12:1930–1935.

18 Gilsanz V, Chalfant J, Mo AO, Lee DC, Dorey FJ, Mittelman SD: Reciprocal relations of subcutaneous and visceral fat to bone structure and strength. J Clin Endocrinol Metab 2009;94:3387–3393.

19 Gabriely I, Ma XH, Yang XM, Atzmon G, Rajala MW, Berg AH, Scherer P, Rossetti L, Barzilai N: Removal of visceral fat prevents insulin resistance and glucose intolerance of aging: an adipokine-mediated process? Diabetes 2002;51:2951–2958.

20 Bjorntorp P: 'Portal' adipose tissue as a generator of risk factors for cardiovascular disease and diabetes. Arteriosclerosis 1990;10:493–496.

21 Boden G: Effects of free fatty acids (FFA) on glucose metabolism: significance for insulin resistance and type 2 diabetes. Exp Clin Endocrinol Diabetes 2003; 111:121–124.

22 Kim SP, Catalano KJ, Hsu IR, Chiu JD, Richey JM, Bergman RN: Nocturnal free fatty acids are uniquely elevated in the longitudinal development of diet-induced insulin resistance and hyperinsulinemia. Am J Physiol Endocrinol Metab 2007;292:E1590–E1598.

23 Nielsen S, Guo Z, Johnson CM, Hensrud DD, Jensen MD: Splanchnic lipolysis in human obesity. J Clin Invest 2004;113:1582–1588.

24 Trayhurn P, Wood IS: Signalling role of adipose tissue: adipokines and inflammation in obesity. Biochem Soc Trans 2005;33:1078–1081.

25 Cartier A, Cote M, Lemieux I, Perusse L, Tremblay A, Bouchard C, Despres JP: Age-related differences in inflammatory markers in men: contribution of visceral adiposity. Metabolism 2009;58:1452–1458.

26 Fontana L, Eagon JC, Trujillo ME, Scherer PE, Klein S: Visceral fat adipokine secretion is associated with systemic inflammation in obese humans. Diabetes 2007;56:1010–1013.

27 Atzmon G, Yang XM, Muzumdar R, Ma XH, Gabriely I, Barzilai N: Differential gene expression between visceral and subcutaneous fat depots. Horm Metab Res 2002;34:622–628.

28 Einstein FH, Atzmon G, Yang XM, Ma XH, Rincon M, Rudin E, Muzumdar R, Barzilai N: Differential responses of visceral and subcutaneous fat depots to nutrients. Diabetes 2005;54:672–678.

29 Giusti V, Suter M, Verdumo C, Gaillard RC, Burckhardt P, Pralong FP: Molecular determinants of human adipose tissue: differences between visceral and subcutaneous compartments in obese women. J Clin Endocrinol Metab 2004;89:1379–1384.

30 van Herpen NA, Schrauwen-Hinderling VB: Lipid accumulation in non-adipose tissue and lipotoxicity. Physiol Behav 2008;94:231–241.

31 Koves TR, Ussher JR, Noland RC, Slentz D, Mosedale M, Ilkayeva O, Bain J, Stevens R, Dyck JR, Newgard CB, Lopaschuk GD, Muoio DM: Mitochondrial overload and incomplete fatty acid oxidation contribute to skeletal muscle insulin resistance. Cell Metab 2008;7:45–56.

32 Coen PM, Dube JJ, Amati F, Stefanovic-Racic M, Ferrell RE, Toledo FG, Goodpaster BH: Insulin resistance is associated with higher intramyocellular triglycerides in type I but not type II myocytes concomitant with higher ceramide content. Diabetes 2010;59:80–88.

33 Pan DA, Lillioja S, Kriketos AD, Milner MR, Baur LA, Bogardus C, Jenkins AB, Storlien LH: Skeletal muscle triglyceride levels are inversely related to insulin action. Diabetes 1997;46:983–88.

34 Kim JK, Fillmore JJ, Chen Y, Yu C, Moore IK, Pypaert M, Lutz EP, Kako Y, Velez-Carrasco W, Goldberg IJ, Breslow JL, Shulman GI: Tissue-specific overexpression of lipoprotein lipase causes tissue-specific insulin resistance. Proc Natl Acad Sci USA 2001;98:7522–7527.

35 Almeda-Valdes P, Cuevas-Ramos D, Aguilar-Salinas CA: Metabolic syndrome and non-alcoholic fatty liver disease. Ann Hepatol 2009;8(suppl 1):S18–S24.

36 Siegel AB, Zhu AX: Metabolic syndrome and hepatocellular carcinoma: two growing epidemics with a potential link. Cancer 2009;115:5651–5661.

37 Korenblat KM, Fabbrini E, Mohammed BS, Klein S: Liver, muscle, and adipose tissue insulin action is directly related to intrahepatic triglyceride content in obese subjects. Gastroenterology 2008;134:1369–1375.

38 Hwang JH, Stein DT, Barzilai N, Cui MH, Tonelli J, Kishore P, Hawkins M: Increased intrahepatic triglyceride is associated with peripheral insulin resistance: in vivo MR imaging and spectroscopy studies. Am J Physiol Endocrinol Metab 2007;293:E1663– E1669.

39 Fabbrini E, Magkos F, Mohammed BS, Pietka T, Abumrad NA, Patterson BW, Okunade A, Klein S: Intrahepatic fat, not visceral fat, is linked with metabolic complications of obesity. Proc Natl Acad Sci USA 2009;106:15430–15435.

40 van der Poorten D, Milner KL, Hui J, Hodge A, Trenell MI, Kench JG, London R, Peduto T, Chisholm DJ, George J: Visceral fat: a key mediator of steatohepatitis in metabolic liver disease. Hepatology 2008;48:449–457.

41 Greenstein AS, Khavandi K, Withers SB, Sonoyama K, Clancy O, Jeziorska M, Laing I, Yates AP, Pemberton PW, Malik RA, Heagerty AM: Local inflammation and hypoxia abolish the protective anticontractile properties of perivascular fat in obese patients. Circulation 2009;119:1661–1670.

42 Carr DB, Utzschneider KM, Hull RL, Kodama K, Retzlaff BM, Brunzell JD, Shofer JB, Fish BE, Knopp RH, Kahn SE: Intra-abdominal fat is a major determinant of the National Cholesterol Education Program Adult Treatment Panel III criteria for the metabolic syndrome. Diabetes 2004;53:2087–2094.

43 Buettner R, Scholmerich J, Bollheimer LC: High-fat diets: modeling the metabolic disorders of human obesity in rodents. Obesity (Silver Spring) 2007;15: 798–808.

44 Powell CS, Blaylock ML, Wang R, Hunter HL, Johanning GL, Nagy TR: Effects of energy expenditure and Ucp1 on photoperiod-induced weight gain in collard lemmings. Obes Res 2002;10:541–550.

45 Davey RA, MacLean HE: Current and future approaches using genetically modified mice in endocrine research. Am J Physiol Endocrinol Metab 2006;291:E429–E438.

46 Friedman JM, Halaas JL: Leptin and the regulation of body weight in mammals. Nature 1998;395:763–770.

47 Masuzaki H, Paterson J, Shinyama H, Morton NM, Mullins JJ, Seckl JR, Flier JS: A transgenic model of visceral obesity and the metabolic syndrome. Science 2001;294:2166–2170.

48 Barzilai N, She L, Liu BQ, Vuguin P, Cohen P, Wang J, Rossetti L: Surgical removal of visceral fat reverses hepatic insulin resistance. Diabetes 1999;48:94–98.

49 Ohman MK, Shen Y, Obimba CI, Wright AP, Warnock M, Lawrence DA, Eitzman DT. Visceral adipose tissue inflammation accelerates atherosclerosis in apolipoprotein E-deficient mice. Circulation 2008;117:798–805.

50 Tran TT, Yamamoto Y, Gesta S, Kahn CR: Beneficial effects of subcutaneous fat transplantation on metabolism. Cell Metab 2008;7:410–420.

51 Einstein FH, Fishman S, Muzumdar RH, Yang XM, Atzmon G, Barzilai N: Accretion of visceral fat and hepatic insulin resistance in pregnant rats. Am J Physiol Endocrinol Metab 2007.

52 Borst SE, Conover CF, Bagby GJ: Association of resistin with visceral fat and muscle insulin resistance. Cytokine 2005;32:39–44.

53 Pitombo C, Araujo EP, De Souza CT, Pareja JC, Geloneze B, Velloso LA: Amelioration of diet-induced diabetes mellitus by removal of visceral fat. J Endocrinol 2006;191:699–706.

54 Shi H, Strader AD, Woods SC, Seeley RJ: The effect of fat removal on glucose tolerance is depot specific in male and female mice. Am J Physiol Endocrinol Metab 2007;293:E1012–E1020.

55 Shi H, Strader AD, Woods SC, Seeley RJ: Sexually dimorphic responses to fat loss after caloric restriction or surgical lipectomy. Am J Physiol Endocrinol Metab 2007;293:E316–E326.

56 Lottati M, Kolka CM, Stefanovski D, Kirkman EL, Bergman RN: Greater omentectomy improves insulin sensitivity in nonobese dogs. Obesity (Silver Spring) 2009;17:674–680.

57 Thorne A, Lonnqvist F, Apelman J, Hellers G, Arner P: A pilot study of long-term effects of a novel obesity treatment: omentectomy in connection with adjustable gastric banding. Int J Obes Relat Metab Disord 2002;26:193–199.

58 Bjorntorp P: Obesity and the risk of cardiovascular disease. Ann Clin Res 1985;17:3–9.

59 Bluher M: The distinction of metabolically 'healthy' from 'unhealthy' obese individuals. Curr Opin Lipidol 2010;21:38–43.

60 Goodpaster BH, Thaete FL, Simoneau JA, Kelley DE: Subcutaneous abdominal fat and thigh muscle composition predict insulin sensitivity independently of visceral fat. Diabetes 1997;46:1579–1585.

61 Porter SA, Massaro JM, Hoffmann U, Vasan RS, O'Donnel CJ, Fox CS: Abdominal subcutaneous adipose tissue: a protective fat depot? Diabetes Care 2009;32:1068–1075.

62 Klein S, Fontana L, Young VL, Coggan AR, Kilo C, Patterson BW, Mohammed BS: Absence of an effect of liposuction on insulin action and risk factors for coronary heart disease. N Engl J Med 2004;350:2549–2557.

63 Mohammed BS, Cohen S, Reeds D, Young VL, Klein S: Long-term effects of large-volume liposuction on metabolic risk factors for coronary heart disease. Obesity (Silver Spring) 2008;16:2648–2651.

64 Busetto L, Bassetto F, Zocchi M, Zuliani F, Nolli ML, Pigozzo S, Coin A, Mazza M, Sergi G, Mazzoleni F, Enzi G: The effects of the surgical removal of subcutaneous adipose tissue on energy expenditure and adipocytokine concentrations in obese women. Nutr Metab Cardiovasc Dis 2008;18:112–120.

65 Ybarra J, Blanco-Vaca F, Fernandez S, Castellvi A, Bonet R, Palomer X, Ordonez-Llanos J, Trius A, Vila-Rovira R, Perez A: The effects of liposuction removal of subcutaneous abdominal fat on lipid metabolism are independent of insulin sensitivity in normal-overweight individuals. Obes Surg 2008;18: 408–414.

66 Kim JY, van de Wall E, Laplante M, Azzara A, Trujillo ME, Hofmann SM, Schraw T, Durand JL, Li H, Li G, Jelicks LA, Mehler MF, Hui DY, Deshaies Y, Shulman GI, Schwartz GJ, Scherer PE: Obesity-associated improvements in metabolic profile through expansion of adipose tissue. J Clin Invest 2007;117:2621–2637.
67 Saltiel AR, Olefsky JM: Thiazolidinediones in the treatment of insulin resistance and type II diabetes. Diabetes 1996;45:1661–1669.
68 Boden G, Cheung P, Mozzoli M, Fried SK: Effect of thiazolidinediones on glucose and fatty acid metabolism in patients with type 2 diabetes. Metabolism 2003;52:753–759.
69 Teranishi T, Ohara T, Maeda K, Zenibayashi M, Kouyama K, Hirota Y, Kawamitsu H, Fujii M, Sugimura K, Kasuga M: Effects of pioglitazone and metformin on intracellular lipid content in liver and skeletal muscle of individuals with type 2 diabetes mellitus. Metabolism 2007;56:1418–1424.
70 Sandouk T, Reda D, Hofmann C: Antidiabetic agent pioglitazone enhances adipocyte differentiation of 3T3-F442A cells. Am J Physiol 1993;264:C1600–C1608.
71 McLaughlin TM, Liu T, Yee G, Abbasi F, Lamendola C, Reaven G, Tsao P, Cushman S, Sherman A: Pioglitazone increases the proportion of small cells in human abdominal subcutaneous adipose tissue. Obesity (Silver Spring) 2009.
72 Adams M, Montague CT, Prins JB, Holder JC, Smith SA, Sanders L, Digby JE, Sewter CP, Lazar MA, Chatterjee VK, O'Rahilly S: Activators of peroxisome proliferator-activated receptor gamma have depot-specific effects on human preadipocyte differentiation. J Clin Invest 1997;100:3149–3153.
73 Bartke A: Insulin and aging. Cell Cycle 2008;7:3338–3343.
74 Berryman DE, List EO, Palmer AJ, Chung MY, Wright-Piekarski J, Lubbers E, O'Connor P, Okada S, Kopchick JJ: Two-year body composition analyses of long-lived GHR null mice. J Gerontol A Biol Sci Med Sci 2010;65:31–40.
75 Rissanen A, Heliovaara M, Knekt P, Aromaa A, Reunanen A, Maatela J: Weight and mortality in Finnish men. J Clin Epidemiol 1989;42:781–789.
76 Lindsted K, Tonstad S, Kuzma JW: Body mass index and patterns of mortality among Seventh-day Adventist men. Int J Obes 1991;15:397–406.
77 Bender R, Jockel KH, Trautner C, Spraul M, Berger M: Effect of age on excess mortality in obesity. JAMA 1999;281:1498–1504.
78 Allison DB, Fontaine KR, Manson JE, Stevens J, VanItallie TB: Annual deaths attributable to obesity in the United States. JAMA 1999;282:1530–1538.
79 Fontaine KR, Redden DT, Wang C, Westfall AO, Allison DB: Years of life lost due to obesity. JAMA 2003;289:187–193.
80 Calle EE, Rodriguez C, Walker-Thurmond K, Thun MJ: Overweight, obesity, and mortality from cancer in a prospectively studied cohort of US adults. N Engl J Med 2003;348:1625–1638.
81 Valdes AM, Andrew T, Gardner JP, Kimura M, Oelsner E, Cherkas LF, Aviv A, Spector TD: Obesity, cigarette smoking, and telomere length in women. Lancet 2005;366:662–664.
82 Stevens J, Keil JE, Rust PF, Tyroler HA, Davis CE, Gazes PC: Body mass index and body girths as predictors of mortality in black and white women. Arch Intern Med 1992;152:1257–1262.
83 Folsom AR, Kaye SA, Sellers TA, Hong CP, Cerhan JR, Potter JD, Prineas RJ: Body fat distribution and 5-year risk of death in older women. JAMA 1993; 269:483–487.
84 Baik I, Ascherio A, Rimm EB, Giovannucci E, Spiegelman D, Stampfer MJ, Willett WC: Adiposity and mortality in men. Am J Epidemiol 2000;152:264–271.
85 Wannamethee SG, Shaper AG, Lennon L, Whincup PH: Decreased muscle mass and increased central adiposity are independently related to mortality in older men. Am J Clin Nutr 2007;86:1339–1346.
86 Pischon T, Boeing H, Hoffmann, et al: General and abdominal adiposity and risk of death in Europe. N Engl J Med 2008;359:2105–2120.
87 Postorino M, Marino C, Tripepi G, Zoccali C: Abdominal obesity and all-cause and cardiovascular mortality in end-stage renal disease. J Am Coll Cardiol 2009;53:1265–1272.
88 Reis JP, Macera CA, Araneta MR, Lindsay SP, Marshall SJ, Wingard DL: Comparison of overall obesity and body fat distribution in predicting risk of mortality. Obesity (Silver Spring) 2009;17:1232–1239.
89 Kuk JL, Ardern CI: Influence of age on the association between various measures of obesity and all-cause mortality. J Am Geriatr Soc 2009.
90 Price GM, Uauy R, Breeze E, Bulpitt CJ, Fletcher AE: Weight, shape, and mortality risk in older persons: elevated waist-hip ratio, not high body mass index, is associated with a greater risk of death. Am J Clin Nutr 2006;84:449–460.
91 Rimm EB, Stampfer MJ, Giovannucci E, Ascherio A, Spiegelman D, Colditz GA, Willett WC: Body size and fat distribution as predictors of coronary heart disease among middle-aged and older US men. Am J Epidemiol 1995;141:1117–1127.
92 Tannenbaum A: The genesis and growth of tumors. II. Effects of caloric restriction per se. Cancer Res 1942;2:460–464.

93 Tannenbaum A: The dependence of the genesis of induced skin tumors on the caloric intake during different stages of carcinogenesis. Cancer Res 1944; 4:673–679.
94 Nutrition Classics: The American Journal of Cancer, Volume XXXVIII, March, 1940: The initiation and growth of tumors. Introduction. I. Effects of underfeeding. By Albert Tannenbaum. Nutr Rev 1987;45: 20–22.
95 Weindruch R, Walford R: The Retardation of Aging and Disease by Dietary Restriction. Springfield, Thomas, 1988.
96 Pugh TD, Klopp RG, Weindruch R: Controlling caloric consumption: protocols for rodents and rhesus monkeys. Neurobiol Aging 1999;20:157–165.
97 Bertrand HA, Lynd FT, Masoro EJ, Yu BP: Changes in adipose mass and cellularity through the adult life of rats fed ad libitum or a life-prolonging restricted diet. J Gerontol 1980;35:827–835.
98 Masoro EJ: Commentary on 'Revisiting the role of fat mass in the life extension induced by caloric restriction'. J Gerontol A Biol Sci Med Sci 1999; 54A:B97.
99 Speakman JR: Body size, energy metabolism and life span. J Exp Biol 2005;208:1717–1730.
100 Bluher M, Kahn BB, Kahn CR: Extended longevity in mice lacking the insulin receptor in adipose tissue. Science 2003;299:572–574.
101 Barzilai N, Gupta G: Revisiting the role of fat mass in the life extension induced by caloric restriction. J Gerontol A Biol Sci Med Sci 1999;54:B89–B96; discussion B97–B98.
102 Harrison DE, Archer JR, Astle CM: Effects of food restriction on aging: separation of food intake and adiposity. Proc Natl Acad Sci USA 1984;81:1835–1838.
103 Berg BN, Simms HS: Nutrition and longevity in the rat. II. Longevity and onset of disease with different levels of food intake. J Nutr 1960;71:255–263.
104 Zhang Y, Proenca R, Maffei M, Barone M, Leopold L, Friedman JM: Positional cloning of the mouse obese gene and its human homologue. Nature 1994; 372:425–432.
105 Frankenberry KA, Somasundar P, McFadden DW, Vona-Davis LC: Leptin induces cell migration and the expression of growth factors in human prostate cancer cells. Am J Surg 2004;188:560–565.
106 Rose DP, Komninou D, Stephenson GD: Obesity, adipocytokines, and insulin resistance in breast cancer. Obes Rev 2004;5:153–165.
107 Barzilai N, Banerjee S, Hawkins M, Chen W, Rossetti L: Caloric restriction reverses hepatic insulin resistance in aging rats by decreasing visceral fat. J Clin Invest 1998;101:1353–1361.
108 Weindruch R: The retardation of aging by caloric restriction: studies in rodents and primates. Toxicol Pathol 1996;24:742–745.
109 Wang C, Weindruch R, Fernandez JR, Coffey CS, Patel P, Allison DB: Caloric restriction and body weight independently affect longevity in Wistar rats. Int J Obes Relat Metab Disord 2004;28:357–362.
110 Muzumdar R, Allison DB, Huffman DM, Ma X, Atzmon G, Einstein FH, Fishman S, Poduval AD, McVei T, Keith SW, Barzilai N: Visceral adipose tissue modulates mammalian longevity. Aging Cell 2008;7:438–440.
111 Kim YW, Kim JY, Lee SK: Surgical removal of visceral fat decreases plasma free fatty acid and increases insulin sensitivity on liver and peripheral tissue in monosodium glutamate (MSG)-obese rats. J Korean Med Sci 1999;14:539–545.

Nir Barzilai, MD
Institute for Aging Research, Departments of Medicine and Molecular Biology
Belfer Building, Suite 701, The Albert Einstein College of Medicine
1300 Morris Park Avenue, Bronx, NY 10461 (USA)
Tel. +1 718 430 3312, Fax +1 718 430 8557, E-Mail barzilai@aecom.yu.edu

Mobbs CV, Hof PR (eds): Body Composition and Aging.
Interdiscipl Top Gerontol. Basel, Karger, 2010, vol 37, pp 20–36

Obesity Paradox during Aging

Ian M. Chapman

Division of Medicine, University of Adelaide, Adelaide, Australia

Abstract

Although obesity in young people is a risk factor for morbidity and mortality, the effect of obesity in the elderly is much more complex. For example, the body weight associated with maximal survival increases with increasing age. Even more striking is the 'obesity paradox' in the elderly, in which overweight is associated with increased risk for cardiovascular disease but decreased mortality from these diseases. Thus, although intentional weight loss by obese older people is probably safe, and likely to be beneficial if they have obesity-related morbidities, caution should be exercised in recommending weight loss to overweight older people on the basis of body weight alone. Methods of achieving weight loss in older adults are the same as in younger adults. Weight loss diets should be combined with an exercise program, if possible, to preserve muscle mass, as dieting results in loss of muscle as well as fat, and older people have reduced skeletal muscle mass compared to younger adults. Weight-loss drugs have not been extensively studied in older people and there is the potential for drug side effects and interactions. Weight loss surgery appears to be safe and effective, although it probably produces less weight loss than in younger adults. Little is yet known about the outcomes of such surgery in people over 65 years.

Prevalence of Overweight and Obesity among Older People

In most developed countries large numbers of older people are overweight. In recent surveys, approximately 71% of Americans 60 years or older and 60% of those 65 years or older were overweight (BMI ≥25), while approximately 32% of those 60 years or older and 20% of those 65 years or older were obese (BMI ≥30) [1, 2]. Similarly, 29% of 55- to 64-year-olds in England were obese [3], while 43% of Australians over 65 were overweight and 25% of 65- to 74-year-olds and 14.4% over 75 years were obese [4].

Not only is obesity common in older people, but it is becoming more common, in line with the increasing rate in younger adults. The prevalence of obesity among people in the USA over 60 years increased from 20 to 32% between 1988–1994 and 1999–2000 [5], and among those over 70 years from 11.4 to 15.5% between 1991 and

2000 [6]. There have been similar increases in other countries [4], including non-western countries. The combined prevalence of overweight and obesity in China more than doubled between 1991 and 1999–2000, from 13 to 27%, with the greatest increases occurring in those older than 65 years [7].

Changes in Body Weight and Body Composition with Increasing Age

An understanding of the causes and consequences of obesity in older people is aided by an understanding of the changes in appetite, food intake, energy expenditure and body composition that occur normally with aging.

Changes in Appetite and Food Intake with Increasing Age

Adults generally become less hungry and eat less as they get older, even if healthy [8]. This physiological, age-related reduction in appetite and energy intake is multi-factorial and has been termed 'the anorexia of aging' [9]. Average daily energy intake decreases by about 30% between 20 and 80 years [8]. Most of this decrease in energy intake is probably a response to the decline in energy expenditure that also occurs with aging. Changes in body weight and body composition reflect the balance of these two declines. Body weight tends to increase through early adult life into middle age, indicating a more rapid decline in energy expenditure than in food intake during this time. In contrast, body weight tends to decrease in older people due to a faster decline in food intake than in energy expenditure in later life.

Changes in Body Weight with Increasing Age

On average, people in westernized countries gain weight until they are about 50–60 years old, remain weight stable for a while and then tend to lose weight [10]. Although some of the decline in body weight after age 50–60 years apparent in cross-sectional studies is due to the premature death of obese people, declining body weight among older people has also been demonstrated in longitudinal studies. In a 2-year, United States, prospective study, community-dwelling men over 65 years lost on average 0.5% of their body weight per year and 13.1% of the group had a weight loss of 4% per annum or more [11]. As a result of weight loss in old age, and the premature death of obese people at younger ages, the prevalence of overweight and obesity, as defined by standard BMI criteria (BMI >25 and ≥30, respectively) peaks around age 55–65 years. It then remains fairly stable until about age 70–75 years, before decreasing [4].

A substantial minority of older people have quite marked weight changes over time [11, 12]. In one study of home-dwelling people over 65 years in the USA [12],

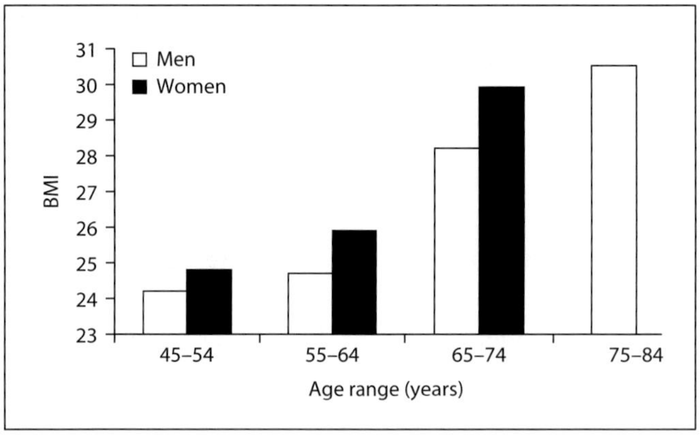

Fig. 1. The age-related increase in BMI associated with 20% increased risk of death relative to people with a BMI of 21 (lowest risk group). Data from the American Cancer Society's Cancer Prevention Study demonstrating a decline in the relative risk of death with increasing body weight with increasing age among 324,135 men and women followed for 12 years. Data not available for women over 75 years. Adapted from Stevens et al. [48].

30% had a greater than 5% change in their body weight over 3 years, with 17% losing more than 5% of their initial body weight and 13% gaining 5% or more. There is evidence that body weight category interacts with changes in body weight to affect health. Adverse effects particularly occur in already underweight people who lose weight and in already overweight people who gain weight [13].

Changes in Body Composition with Increasing Age

Enlargement and Redistribution of Fat Stores

With aging there is usually a steady increase in fat stores and decrease in fat-free mass, the latter mainly due to loss of skeletal muscle. Consequently, at any given weight or body mass index, older people have, on average, substantially more body fat than young adults. In one study, the mean body fat percentage of 75-year-old men weighing 80 kg was 29%, compared to 15% in 20-year-old men of the same weight [14] (fig. 1). The increase in body fat with aging is multifactorial in origin, with decreased physical activity being a cause, as well as contributions from reduced growth hormone secretion, declining sex hormone action and reduced resting metabolic rate and thermic effect of food.

Older adults not only have more body fat than young adults, but their fat is distributed differently. A greater proportion of body fat in older than young people is intrahepatic, intramuscular, and intra-abdominal (versus subcutaneous) [15, 16], changes that in both young and older adults are associated with increased insulin resistance [16]. In one study, intramuscular fat stores were 50% greater, intrahepatic stores four times greater and insulin resistance two times greater in older (65–74 years) than young adults (20–32 years) [16]. In younger adults such changes to body fat stores and increases in insulin resistance are associated with adverse metabolic outcomes, including increased rates of diabetes mellitus and cardiovascular disease. While it might, therefore, be predicted that the age-related changes in body fat stores would lead to bad metabolic outcomes in older people, this is not proven. Indeed, given that the body weight compatible with longest survival increases with increasing age (see below), and much if not all of the increase in body weight is due to increased fat stores, it may be that advancing age in some way blunts the harmful effects of increasing body fat.

Loss of Skeletal Muscle (Sarcopenia)

Aging is associated with a decrease in muscle mass and strength, with a loss of up to 3 kg of lean body mass per decade after age 50 years. After age 60 years, loss of body weight is disproportionately of lean body tissue, predominantly skeletal muscle. The causes of age-related skeletal muscle loss are multiple and not fully understood, but probably similar to those leading to fat gain, including reduced exercise and anabolic hormone action. When excessive, this loss of skeletal muscle leads to sarcopenia (from the Greek meaning 'poverty of flesh'). Sarcopenia has been defined in various ways, including as a skeletal mass more than 2 SDs below the young adult sex-specific mean [17]. The prevalence of sarcopenia in older people varies according to the population studied and diagnostic criteria used, but is in the range of 6–15% in people over 65 years [17]. The prevalence of sarcopenia increases dramatically as the 'young-old' age, and is up to four times higher in those over 85 years than in those 70–75 years [18].

The reduction of skeletal muscle mass and strength in sarcopenia is so severe that it is often associated with marked functional impairment. The presence of sarcopenia is an independent predictor of poor gait, balance, falls, and fractures. In the NHANES III study, for example, older people with marked sarcopenia (less than 5.75 kg skeletal muscle/m^2) were 3.3 times (women) to 4.7 times (men) more likely to have physical disability than those with low-risk skeletal muscle mass (more than 6.75 kg/m^2) [19].

Sarcopenic Obesity

In young adults, obesity tends to be associated with increased skeletal muscle, acquired to support the extra weight. In contrast, in older people, excess weight, to the point of obesity, can co-exist with muscle loss and even sarcopenia. Hence, the existence of sarcopenic obesity, or the 'skinny fat' elderly, who have high levels of body fat but do not look overweight as they have lost much of their lean tissue. This combination in people with sarcopenic obesity of an excess of (possibly) metabolically bad fat tissue and a deficiency of beneficial muscle is associated with particularly adverse effects [20]. In one prospective study of elderly people followed for up to 8 years, those with sarcopenic obesity (skeletal muscle mass more than 2 SDs below young adult mean AND percentage body fat above the 60th percentile) at baseline were 2–3 times more likely to develop disabilities in activities of daily living than were lean sarcopaenic, nonsarcopenic obese or normal body composition subjects [21]. The age-related loss of skeletal muscle and its adverse effects probably helps explain the consistently demonstrated benefits of exercise programs in elderly people, particularly those that increase muscle mass and function.

Causes of Overweight and Obesity in Older People

As indicated above, epidemiological studies show that over the last few decades people have not usually gained weight during old age. It follows from this that the high and increasing proportion of overweight and obesity in older people observed over recent decades is mainly due to the high and increasing proportion of adults who reach old age already overweight. The causes of obesity in the elderly are therefore largely those of obesity in younger adults. These are complex, multifactorial and relate to changing environmental factors which have made it easier for people to consume more energy than they expend [22]. They have been extensively reviewed elsewhere and will not be addressed further in this chapter.

Older people are often not weight stable and a substantial minority of older people have quite marked weight changes over time [11, 12]. There are therefore some people who only become overweight or whose weight increases dramatically in later life. This can be due to medical illnesses, which restrict mobility and hence alter the food intake-energy expenditure balance in favor of the former. Most common among these is musculoskeletal disease, such as osteoarthritis of the lower limbs or back, while cardiac failure and depression are also important causes, as is increased food intake due to corticosteroid use.

Consequences of Obesity in Older People

As in younger adults, obesity in older adults is associated with absolute and relative increases in both mortality and morbidity. Obesity-related co-morbidities are summarized in table 1. Most of these conditions become more common and severe with increasing age even in those who are not overweight, but their adverse effects are exacerbated by excess weight.

Several apparent paradoxes related to obesity in older people will be considered below. Firstly, obesity in older people appears to be associated with the same if not greater morbidities and reduction of quality of life as in younger adults, but the association of excess weight with increased mortality is substantially weaker than in younger people. In the very old there may, indeed, be no association between excess

Table 1. Obesity-related comorbidities in older people

Reduced mobility
Reduced quality of life
Type 2 diabetes mellitus
Hypertension
Cardiovascular disease
Sleep apnea
Increased rate of malignancies
Breast
Colon
Prostate
Uterus
Fatty liver
Thromboembolic disease

body weight and reduced life expectancy, except perhaps in the extremely obese. Secondly, the effects of weight loss by 'overweight' older people are not well defined and may be adverse, even if weight loss is intentional. For example, even though obesity is associated with increased cardiovascular disease, obesity also reduces the mortality rate from such diseases (see below).

Morbidity, Function and Quality of Life

As in younger adults, obesity in older people is associated with increased rates of cataracts, mechanical urinary and bladder problems, sleep apnea and other respiratory problems (reviewed in [10]), vascular disease, hypertension and type 2 diabetes mellitus [23]. After the age of 65 years, over 60% of people have symptomatic osteoarthritis, commonly affecting the hip and knee, and this is a major cause of disability. Excess weight hastens the development of osteoarthritis of the knee and possibly other joints [24, 25], exacerbates the symptoms of lower limb osteoarthritis, and makes surgical treatment more hazardous.

Functional capacity and mobility are significantly reduced in obese compared to lean older adults [26]. The association of obesity and functional disability is particularly strong in older adults, in whom it is associated with reduced muscle mass and strength and hence with physical frailty [20]. Obese older people have increased difficulties performing activities of daily living, including dressing and getting in and out of bed [27, 28]. Compared to their nonobese counterparts, obese older people are less likely to be pain free, have greater limitations of physical function, and are more likely to be homebound. Obesity is predictive of a greater rate of future disability, declines in functional status [10], particularly when associated with loss of skeletal mass [21], and an increased admission rate to nursing homes [29]. Among women age 65 years or more in the Nurses Health Study, a weight gain of 20 lb (≈9 kg) or more over 4 years was associated with a reduction in reported physical functioning [13]. Increased fat mass appears to be the factor specifically responsible for obesity-related disability [26].

Obesity is associated with a reduced quality of life [30], due to multiple factors, including associated joint pain [31]. There have been reports that overweight women have a greater disease burden and greater reduction in quality of life than overweight men [27, 32]. Obesity has been linked with depression, but study results are conflicting [33–35]. There is evidence for link between obesity and the later development of Alzheimer type dementia, but evidence for this is not clear cut, and may be confused by the associated conditions of diabetes and hypertension, both of which are linked to dementia [36].

Obesity, particularly central obesity, is a risk factor for and component of the metabolic syndrome. It is associated with and predisposes to hypertension, dylipidemia and cardiovascular disease among adults of all ages. There is an increased rate of vascular disease, particularly ischemic heart disease, congestive cardiac failure and cerebrovascular disease in overweight people, including older people [37].

The Obesity Paradox in the Elderly

Although obesity is a risk factor for the development of vascular disease, it is now well established that when obese people develop these vascular diseases they have better outcomes than their normal weight counterparts. This has been termed the 'obesity paradox', and applies for people with hypertension, coronary artery disease, congestive cardiac failure, peripheral arterial disease and those referred for exercise testing [38–42]. Studies have consistently shown reductions in mortality of up to 40% in people with heart failure who are obese, compared to individuals without elevated BMIs [38]. In one study of over 108,927 subjects with de-compensated heart failure, mortality decreased by 10% for every 5-unit increase in BMI [43]. This paradox has attracted considerable interest, and a number of potential mechanisms for the protective effects of body fat and increased BMI have been proposed. These include greater metabolic reserve, protective effects of certain fat-derived factors such as cytokines and of higher blood pressure, reduced activation of the sympathetic nervous system, and lower circulating concentrations of natriuretic peptides [38].

Preservation of Bone Mass

A beneficial effect of obesity in older people is its association with increased bone density. This combines with cushioning of falls provided by the extra fat stores, particularly around the hips ('endogenous' hip protectors) to reduce fracture rates in older people [44]. Most studies show that when overweight adults of any age intentionally lose weight they also lose bone [10, 45]. Substantial unintentional weight loss in older people is associated with an increased risk of hip fracture [46]. This is probably due to a combination of reduced bone mass and the effects of whatever illness caused the weight loss, for example by increasing the likelihood of falls. Although it is reasonable to assume that there is at least some increase in fracture risk due to the weight loss-associated bone loss, in overweight older people who intentionally lose weight, this has not been established.

Mortality

The harmful effects of overweight/obesity on life expectancy are weaker in older than young adults (see below). Nevertheless, the relative risk of mortality is increased at high BMIs until about the age of 75 years, and because of the greater background death rate in older people the absolute increase in death rate attributable to obesity is substantial; 25% of the excess deaths attributed to obesity in a NHANES analysis occurred in people over 70 years [5]. The causes of increased mortality are essentially the same as those in younger adults: diabetes, hypertension, sleep apnea, cardiovascular disease and an increased risk with obesity of developing certain cancers, including breast, uterus, colon and prostate.

The relative increase in the risk of death associated with being obese compared to being normal weight is not as great in older as in young adults. An assessment of 13 observational, prospective studies in which nonhospitalized people over 65 years were followed for at least 3 years [47], found an association between mortality and increased BMI in only a few, and then only above a BMI of 27–28.5, with little or no increase in mortality at any BMI for people over 75 years. Where an optimum BMI could be identified it was usually in the range 27–30. Consistent with this, in the large, prospective, American Cancer Society's Cancer Prevention Study, involving over 300,000 adults, the BMI associated with a 20% relative increase in mortality above the optimum BMI of 21 rose steadily with age, from 23.8 at 30–44 years to 30.5 at 75–84 years [48] (fig. 1). In the same study, the relative risk of death associated with every increment of 1.0 in BMI decreased significantly with increasing age between age 30 years and ≥85 years, such that there was no significant increase in relative risk of death above age 75 years. Similarly, a combined analysis of the American NHANES I-III (1974–2000) study results found no significant increase in mortality with any degree of overweight in people over 70 years, and an increased death rate only in those with a BMI ≥35 among those 60–69 years [49]. There is also evidence that, compared to other weight groups, overweight and obese people with end-stage renal failure on dialysis, chronic obstructive airways disease, HIV/AIDS and certain cancers, as well as after vascular and other surgery, have lower mortality than their normal-weight counterparts [22, 38, 50].

Of note, low as well as high body weights are associated with increased mortality in adults of all ages, including the elderly. The relationship between BMI and mortality is U-shaped, with a clear increase in mortality in people whose BMI is less than 18.5–19 [49, 51].

The above findings of a weakening association between overweight/obesity and increased mortality with increasing age, relate to BMI, not to the amount of body fat. On average, older people have substantially more body fat than young people with the same BMI [14]. Thus, the positive association between elevated amounts of body fat and mortality present in young adults weakens with increasing age even more than that between increasing BMI and mortality. This raises the possibility that fat stores in old people are different to and metabolically 'better' than fat stores in younger adults. As indicated above [14–16], there are indeed differences between the type and distribution of fat stores in older compared to young adults, and these may underpin some of these differences in effect.

Another interesting point is that the adverse effects of obesity on life expectancy may have weakened over recent decades, in all age groups including the elderly, possibly due to improved medical care for the associated metabolic ill effects such as hypertension and type 2 diabetes mellitus. In the NHANES studies between 1971–1975 and 1988–1994, there was a decrease of 63% in the mortality rates associated with obesity in the two later studies compared to the first [49].

Management of Obesity in Older People

Should Overweight Older People Be Advised to Lose Weight?

This is a controversial question. Obesity in older people is associated with increased morbidity and reduced life expectancy at very high BMIs up to at least 70 years. There is evidence that weight loss by overweight, older people is associated with improved quality of life. In the Nurses' Health Study, weight loss in initially overweight women was associated with improved physical function and vitality as well as decreased bodily pain [13]. In a small study of 27 older (mean age 70 years), obese subjects randomized to 6 months of control therapy or behavioral therapy for weight loss plus exercise training three times per week, the active treatment group lost 8.4% of their body weight and had significant improvements in physical performance scores, peak oxygen consumption and functional status questionnaire scores, compared to the control subjects, whose weight did not change [52]. Of note, the actively treated group lost fat mass, but had no decrease in fat-free mass. Other studies have also shown significant improvements in health related quality of life in adults with type 2 diabetes who lose substantial amounts of weight, irrespective of their age [53], and substantial improvements in quality of life in people over age 60 years who lost substantial amounts of weight after bariatric surgery [54].

The apparently beneficial effects of weight loss for obese older people must be balanced against the detrimental effects of weight loss on muscle mass and bone density and the association, repeatedly detected in large population studies, between all-cause weight loss and increased mortality in older people, even those who are initially overweight [55–60]. In the Rancho Bernardo Study of 1,801 older men and women, weight loss was associated with a significant 38–76% increase in mortality in those without diabetes, and a 65–366% increase in mortality in those with diabetes [59]. In an intriguing study of 1,114 Finnish men, sequentially examined at mean ages 25, 47 (mid-life) and 73 years (late-life), and classified as to whether they were normal weight (BMI ≤25) or overweight (BMI >25) at these time points, the highest mortality was in the men who were overweight in mid-life and then lost weight to be normal weight in late life (2 times greater than in both the constantly normal weight group and in the group normal weight in mid-life and overweight in late life, $p < 0.001$) [42].

The increased mortality associated with weight loss in older people appears related to whether such weight loss is intentional or unintentional, the latter being a marker of underlying diseases such as cancer. It is probably also related to how weight loss is achieved. There is good evidence that older people do better when physical exercise is a major component of their weight loss program.

Effects of Intentional versus Unintentional Weight Loss

Studies demonstrating an association between weight loss and increased mortality in older people have largely examined all-cause weight loss, whether intentional or unintentional. There is little doubt that unintentional weight loss is not good for the elderly. Although some study results have been interpreted to show increased mortality after even intentional weight loss in older people [12], it is difficult to determine what proportion of weight loss labeled intentional was instead unintentional. In a meta-analysis of 26 studies examining the connection between weight loss and mortality [55], weight loss which was unintentional or ill-defined was associated with a significant 22–39% increase in mortality, whereas intentional weight loss had no significant effect on mortality. Intentional weight loss had a small benefit on mortality for unhealthy people (those with obesity-related risk factors) and even more for the unhealthy obese (RR 0.84 [95% CI 0.73–0.97, $p = 0.018$]), whereas it was associated with slightly increased mortality in healthy individuals (RR 1.11 [95% CI 1–1.22, $p = 0.05$]). On balance, this and other studies [61] suggest that intentional weight loss by initially overweight older people, particularly if they have features of the metabolic syndrome, has either no significant effect or even beneficial effects on mortality. In the US National Health Interview Survey, which followed 20,847 adults with mean age 54 years for 9 years, all-cause weight loss was associated with a significant increase in mortality, as in other studies [62]. Reported attempted weight loss, however, even if unsuccessful, was associated with a 24% reduction in mortality (RR 0.76 [0.64–0.9]), as was successful intentional weight loss (RR 0.76 [0.6–0.97]). There was no interaction between weight loss intention and age in the effect on mortality, consistent with a prolongation of life by intended weight loss in older as well as young adults.

Studies involving predominantly younger adults have found that intentional weight loss can reduce mortality in those with obesity-related health problems, such as type 2 diabetes, ischemic heart disease and hypertension [63]. Although such studies have not been done in older people, available evidence provides no evidence of harm and suggests that it is safe to recommend weight loss to overweight older people with obesity-related morbidities, particularly reduced mobility and function. This is the group that appears to have the most to gain. There are few if any indications for recommending weight loss to older people based on their weight alone.

Effects of Losing Fat versus Losing Lean Tissue in Older People

Part of the apparent discrepancy in overweight, older people between the beneficial effects of weight loss on morbidity and quality of life on the one hand, and the increased mortality often associated with weight loss on the other, may relate to the particular components of body tissue that make up the weight loss. A substantial amount of the tissue lost when older people lose weight is lean tissue, i.e. muscle. Any loss of this is likely to be harmful, whereas loss of fat tissue seems likely to be more beneficial. This is supported by the findings of Allison et al. [57], who examined the independent effects of weight loss and fat loss on all cause mortality in two prospective studies of 4,621 subjects. In both studies weight loss was associated with increased, and fat loss with decreased, mortality. This emphasizes the importance in older people, who have reduced lean tissue stores (muscle) and suffer more than young adults from its loss, to preserve lean tissue if possible and only lose fat tissue when weight is intentionally lost.

Weight Loss Measures for Overweight Older People

As indicated above, it may not be appropriate for older overweight people to try to lose weight. When it is, due to obesity-related morbidity and reduced quality of life, the treatment options to achieve weight loss are the same as those in younger adults; changes in diet and exercise, medications and surgery [10]. There is limited, although

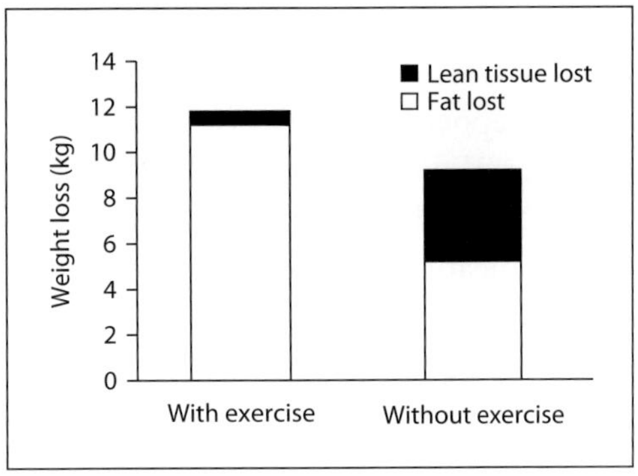

Fig. 2. Muscle-preserving effect of exercise during a weight loss diet: the effect on type of tissue lost during an 8-week weight loss program, in 72 mildly obese men, randomly assigned to exercise or no exercise. Adapted from Prentice and Jebb [14] and Pavlou et al. [66].

growing, information about the effectiveness and safety of weight loss treatments, particularly medications and surgery, in older people.

Lifestyle Measures

Lifestyle interventions, combining a reduced energy diet (reduction of about 500–750 kcal/day) and exercise are at least as effective in producing weight loss in people over 60 years as in younger adults [10], and possibly more so [64]. The incorporation of an exercise component, particularly weight-bearing aerobic and resistance exercise, is very important, as low muscle mass (even sarcopenia) is often present even before deliberate weight loss, and is usually exacerbated by it. In the absence of an exercise program about 25%, and sometimes considerably more, of the diet-induced weight loss is accounted for by loss of lean tissue, essentially skeletal muscle [65, 66]. The addition of an exercise program to a weight loss diet reduces this loss of lean tissue, often dramatically (fig. 2). A meta-analysis of the results of 28 studies, not specifically performed in older adults, found that the addition of an exercise program to diet reduces the diet-induced loss of muscle significantly, from 29 to 17% of weight lost in men and from 22 to 17% in women [65]. Villareal et al. [52] found that obese subjects, 65 years or older, who combined three 90 min exercise sessions per week with a supervised weight loss diet were able to lose 8.4% of their body weight over 6 months without loss of fat-free mass (muscle) and with significant increases in muscle strength. Other studies in older people have shown that exercise inhibits the regional loss of bone density that accompanies weight loss, improves physical function by increasing muscle mass and fitness [67, 68], and reduces the risk of falls [69]. Older people should be assessed prior to starting an exercise program, to determine whether there is a need for formal stress testing to exclude significant underlying coronary artery disease and the exercise intensity should be gradually increased. Multivitamin supplements together with calcium (1,000–1,500 mg/day) and vitamin D (800–1,000 IU/day) for bone protection should also be taken.

Medications

There is little reported experience with weight loss drugs such as sibutramine, which reduces appetite and food intake, and orlistat, which inhibits lipase activity and thus fat absorption, in older people. Such drugs should be used with caution in older people, because of limited efficacy data, the possibility of interactions with other (multiple) medications, and potentially worse side effects. Orlistat appears to be as effective in older as in young adults [10], with a weight loss over 1 year of approximately 3 kg more than with placebo, but can cause diarrhea and other gastrointestinal side effects such as fecal incontinence. Side effects of sibutramine can include insomnia, constipation and increased blood pressure.

Surgery for Weight Loss

Only a few studies so far have examined the outcomes of bariatric surgery in people over 60 years. The results are encouraging [70]. In three studies comparing the effects of gastric bypass surgery in 115 people over 60 years to those in 3,470 people less than 60 years [71–73], there was one peri-operative death in the older group (0.86%) and a possible but not definite slight increase in operative complication rate compared to the younger patients. The surgery resulted in a marked mean weight loss of 39–43 kg at 1 year in the older subjects and a significant reduction in the number of obesity-related morbidities and number of medications needed to treat these. Both reductions were slightly, but significantly less in older than young adults. These results were supported by those of Busetto et al. [74], who compared the outcomes of laparoscopic banding surgery in 216 patients ≥60 years (mean age 64.1 years) to those in 5,274 patients aged less than 60 years. They found similar surgical outcomes in the two groups, a slightly higher peri-operative mortality in the older group (0.46 vs. 0.04%), and comparable substantial reductions in morbidities, particularly diabetes, sleep apnea and dyslipidemia, despite a significantly lower reduction in BMI among the older patients (≈12 vs. 20% weight reduction at 1 year, $p < 0.05$). The mean age of the 'older' subjects in these studies was quite young, however, at between 63–65 years, and the results of such surgery in even older people remain to be determined. The numbers of surgically treated older people in published studies are still relatively low, and caution should be exercised in recommending bariatric surgery to particularly the very old. Nevertheless, available evidence suggests that bariatric surgery is an effective weight loss option in older people, particularly for selected older people substantially disabled by complications of obesity.

Conclusions

Many older people are overweight or obese and this is associated with increased morbidity and reduced quality of life. The adverse effects of obesity on life expectancy in older adults are less dramatic than in young adults, particularly among the very old. Weight loss is achievable and is often beneficial in those people in whom excess weight is associated with functional impairments. Weight loss should not be recommended to older people on the basis of their body weight alone, rather than co-existent morbidity. Older people have less lean tissue (muscle) than younger adults and can be harmed by further loss of lean tissue during diet-induced weight loss. Wherever possible, therefore, weight loss measures should aim to preserve lean tissue by including an exercise component. Bariatric surgery is used increasingly for the treatment of older, obese people. Results to date indicate acceptable safety and comparable reductions in morbidity and improvements in quality of life, despite lesser loss of weight, compared to younger adults undergoing these procedures.

References

1 Li F, Fisher KJ, Harmer P: Prevalence of overweight and obesity in older U.S. adults: estimates from the 2003 Behavioral Risk Factor Surveillance System survey. J Am Geriatr Soc 2005;53:737–739.

2 Ogden CL, Carroll MD, Curtin LR, McDowell MA, Tabak CJ, Flegal KM: Prevalence of overweight and obesity in the United States, 1999–2004. JAMA 2006;295:1549–1555.

3 Banks J, Marmot M, Oldfield Z, Smith JP: Disease and disadvantage in the United States and in England. JAMA 2006;295:2037–2045.

4 Cameron AJ, Welborn TA, Zimmet PZ, Dunstan DW, Owen N, Salmon J, Dalton M, Jolley D, Shaw JE: Overweight and obesity in Australia: the 1999–2000 Australian Diabetes, Obesity and Lifestyle Study (AusDiab). Med J Aust 2003;178:427–432.

5 Flegal KM, Carroll MD, Ogden CL, Johnson CL: Prevalence and trends in obesity among US adults, 1999–2000. JAMA 2002;288:1723–1727.

6 Mokdad AH, Bowman BA, Ford ES, Vinicor F, Marks JS, Koplan JP: The continuing epidemics of obesity and diabetes in the United States. JAMA 2001;286:1195–1200.

7 Wildman RP, Gu D, Muntner P, Wu X, Reynolds K, Duan X, Chen CS, Huang G, Bazzano LA, He J: Trends in overweight and obesity in Chinese adults: between 1991 and 1999–2000. Obesity (Silver Spring) 2008;16:1448–1453.

8 Wurtman JJ, Lieberman H, Tsay R, Nader T, Chew B: Calorie and nutrient intakes of elderly and young subjects measured under identical conditions. J Gerontol 1988;43:B174–B180.

9 Morley JE: Anorexia of aging: physiologic and pathologic. Am J Clin Nutr 1997;66:760–773.

10 Villareal DT, Apovian CM, Kushner RF, Klein S: Obesity in older adults: technical review and position statement of the American Society for Nutrition and NAASO, The Obesity Society. Am J Clin Nutr 2005;82:923–934.

11 Wallace JI, Schwartz RS, LaCroix AZ, Uhlmann RF, Pearlman RA: Involuntary weight loss in older outpatients: incidence and clinical significance. J Am Geriatr Soc 1995;43:329–337.

12 Fried LP, Tangen CM, Walston J, Newman AB, Hirsch C, Gottdiener J, Seeman T, Tracy R, Kop WJ, Burke G, McBurnie MA: Frailty in older adults: evidence for a phenotype. J Gerontol A Biol Sci Med Sci 2001;56:M146–M156.

13 Fine JT, Colditz GA, Coakley EH, Moseley G, Manson JE, Willett WC, Kawachi I: A prospective study of weight change and health-related quality of life in women. JAMA 1999;282:2136–2142.

14 Prentice AM, Jebb SA: Beyond body mass index. Obes Rev 2001;2:141–147.

15 Beaufrere B, Morio B: Fat and protein redistribution with aging: metabolic considerations. Eur J Clin Nutr 2000;54(suppl 3):S48–S53.

16 Cree MG, Newcomer BR, Katsanos CS, Sheffield-Moore M, Chinkes D, Aarsland A, Urban R, Wolfe RR: Intramuscular and liver triglycerides are increased in the elderly. J Clin Endocrinol Metab 2004;89: 3864–3871.

17 Melton LJ 3rd, Khosla S, Riggs BL: Epidemiology of sarcopenia. Mayo Clin Proc 2000;75(suppl):S10–S12; discussion S2–S3.

18 Castillo EM, Goodman-Gruen D, Kritz-Silverstein D, Morton DJ, Wingard DL, Barrett-Connor E: Sarcopenia in elderly men and women: the Rancho Bernardo study. Am J Prev Med 2003;25:226–231.

19 Janssen I, Baumgartner RN, Ross R, Rosenberg IH, Roubenoff R: Skeletal muscle cutpoints associated with elevated physical disability risk in older men and women. Am J Epidemiol 2004;159:413–421.
20 Villareal DT, Banks M, Siener C, Sinacore DR, Klein S: Physical frailty and body composition in obese elderly men and women. Obes Res 2004;12:913–920.
21 Baumgartner RN, Wayne SJ, Waters DL, Janssen I, Gallagher D, Morley JE: Sarcopenic obesity predicts instrumental activities of daily living disability in the elderly. Obes Res 2004;12:1995–2004.
22 Davenport DL, Xenos ES, Hosokawa P, Radford J, Henderson WG, Endean ED: The influence of body mass index obesity status on vascular surgery 30-day morbidity and mortality. J Vasc Surg 2009;49:140–147, 7 e1; discussion 7.
23 Folsom AR, Kushi LH, Anderson KE, Mink PJ, Olson JE, Hong CP, Sellers TA, Lazovich D, Prineas RJ: Associations of general and abdominal obesity with multiple health outcomes in older women: the Iowa Women's Health Study. Arch Intern Med 2000; 160:2117–2128.
24 Cicuttini FM, Baker JR, Spector TD: The association of obesity with osteoarthritis of the hand and knee in women: a twin study. J Rheumatol 1996;23: 1221–1226.
25 Grotle M, Hagen KB, Natvig B, Dahl FA, Kvien TK: Obesity and osteoarthritis in knee, hip and/or hand: an epidemiological study in the general population with 10 years follow-up. BMC Musculoskelet Disord 2008;9:132.
26 Jensen GL: Obesity and functional decline: epidemiology and geriatric consequences. Clin Geriatr Med 2005;21:677–687, v.
27 Peytremann-Bridevaux I, Santos-Eggimann B: Health correlates of overweight and obesity in adults aged 50 years and over: results from the Survey of Health, Ageing and Retirement in Europe (SHARE). Obesity and health in Europeans aged > or = 50 years. Swiss Med Wkly 2008;138:261–266.
28 Lang IA, Llewellyn DJ, Alexander K, Melzer D: Obesity, physical function, and mortality in older adults. J Am Geriatr Soc 2008;56:1474–1478.
29 Zizza CA, Herring A, Stevens J, Popkin BM: Obesity affects nursing-care facility admission among whites but not blacks. Obes Res 2002;10:816–823.
30 Jia H, Lubetkin EI: The impact of obesity on health-related quality-of-life in the general adult US population. J Public Health (Oxf) 2005;27:156–164.
31 Heo M, Allison DB, Faith MS, Zhu S, Fontaine KR: Obesity and quality of life: mediating effects of pain and comorbidities. Obes Res 2003;11:209–216.
32 Muennig P, Lubetkin E, Jia H, Franks P: Gender and the burden of disease attributable to obesity. Am J Publ Health 2006;96:1662–1668.
33 Simon GE, Ludman EJ, Linde JA, Operskalski BH, Ichikawa L, Rohde P, Finch EA, Jeffery RW: Association between obesity and depression in middle-aged women. Gen Hosp Psychiatry 2008;30:32–39.
34 Ho RC, Niti M, Kua EH, Ng TP: Body mass index, waist circumference, waist-hip ratio and depressive symptoms in Chinese elderly: a population-based study. Int J Geriatr Psychiatry 2008;23:401–408.
35 Blazer DG, Moody-Ayers S, Craft-Morgan J, Burchett B: Depression in diabetes and obesity: racial/ethnic/gender issues in older adults. J Psychosom Res 2002;53:913–916.
36 Salihu HM, Bonnema SM, Alio AP: Obesity: What is an elderly population growing into? Maturitas 2009;63:7–12.
37 Darmon P, Kaiser MJ, Bauer JM, Sieber CC, Pichard C: Restrictive diets in the elderly: Never say never again? Clin Nutr 2010;29:170–174.
38 Lavie CJ, Milani RV, Ventura HO: Obesity and cardiovascular disease: risk factor, paradox, and impact of weight loss. J Am Coll Cardiol 2009;53:1925–1932.
39 McAuley P, Myers J, Abella J, Froelicher V: Body mass, fitness and survival in veteran patients: another obesity paradox? Am J Med 2007;120:518–524.
40 Uretsky S, Messerli FH, Bangalore S, Champion A, Cooper-Dehoff RM, Zhou Q, Pepine CJ: Obesity paradox in patients with hypertension and coronary artery disease. Am J Med 2007;120:863–870.
41 Galal W, van Gestel YR, Hoeks SE, Sin DD, Winkel TA, Bax JJ, Verhagen H, Awara AM, Klein J, van Domburg RT, Poldermans D: The obesity paradox in patients with peripheral arterial disease. Chest 2008;134:925–930.
42 Strandberg TE, Strandberg AY, Salomaa VV, Pitkala KH, Tilvis RS, Sirola J, Miettinen TA: Explaining the obesity paradox: cardiovascular risk, weight change, and mortality during long-term follow-up in men. Eur Heart J 2009;30:1720–1727.
43 Fonarow GC, Srikanthan P, Costanzo MR, Cintron GB, Lopatin M: An obesity paradox in acute heart failure: analysis of body mass index and inhospital mortality for 108,927 patients in the Acute Decompensated Heart Failure National Registry. Am Heart J 2007;153:74–81.
44 Schott AM, Cormier C, Hans D, Favier F, Hausherr E, Dargent-Molina P, Delmas PD, Ribot C, Sebert JL, Breart G, Meunier PJ: How hip and whole-body bone mineral density predict hip fracture in elderly women: the EPIDOS Prospective Study. Osteoporos Int 1998;8:247–254.

45 Ensrud KE, Fullman RL, Barrett-Connor E, Cauley JA, Stefanick ML, Fink HA, Lewis C E, Orwoll E: Voluntary weight reduction in older men increases hip bone loss: the osteoporotic fractures in men study. J Clin Endocrinol Metab 2005;90:1998–2004.
46 Langlois JA, Harris T, Looker AC, Madans J: Weight change between age 50 years and old age is associated with risk of hip fracture in white women aged 67 years and older. Arch Intern Med 1996;156:989–994.
47 Heiat A, Vaccarino V, Krumholz HM: An evidence-based assessment of federal guidelines for overweight and obesity as they apply to elderly persons. Arch Intern Med 2001;161:1194–1203.
48 Stevens J, Cai J, Pamuk ER, Williamson DF, Thun MJ, Wood JL: The effect of age on the association between body-mass index and mortality. N Engl J Med 1998;338:1–7.
49 Flegal KM, Graubard BI, Williamson DF, Gail MH: Excess deaths associated with underweight, overweight, and obesity. JAMA 2005;293:1861–1867.
50 Mullen JT, Moorman DW, Davenport DL: The obesity paradox: body mass index and outcomes in patients undergoing nonbariatric general surgery. Ann Surg 2009;250:166–172.
51 Willett WC, Dietz WH, Colditz GA: Guidelines for healthy weight. N Engl J Med 1999;341:427–434.
52 Villareal DT, Banks M, Sinacore DR, Siener C, Klein S: Effect of weight loss and exercise on frailty in obese older adults. Arch Intern Med 2006;166:860–866.
53 Williamson DA, Rejeski J, Lang W, Van Dorsten B, Fabricatore AN, Toledo K: Impact of a weight management program on health-related quality of life in overweight adults with type 2 diabetes. Arch Intern Med 2009;169:163–171.
54 Taylor CJ, Layani L: Laparoscopic adjustable gastric banding in patients > or = 60 years old: is it worthwhile? Obes Surg 2006;16:1579–1583.
55 Harrington M, Gibson S, Cottrell RC: A review and meta-analysis of the effect of weight loss on all-cause mortality risk. Nutr Res Rev 2009;22:93–108.
56 Andres R, Muller DC, Sorkin JD: Long-term effects of change in body weight on all-cause mortality: a review. Ann Intern Med 1993;119(7 Pt 2):737–743.
57 Allison DB, Zannolli R, Faith MS, Heo M, Pietrobelli A, VanItallie TB, Pi-Sunyer FX, Heymsfield SB: Weight loss increases and fat loss decreases all-cause mortality rate: results from two independent cohort studies. Int J Obes Relat Metab Disord 1999;23:603–611.
58 Seidell JC, Visscher TL: Body weight and weight change and their health implications for the elderly. Eur J Clin Nutr 2000;54(suppl 3):S33–S39.
59 Wedick NM, Barrett-Connor E, Knoke JD, Wingard DL: The relationship between weight loss and all-cause mortality in older men and women with and without diabetes mellitus: the Rancho Bernardo study. J Am Geriatr Soc 2002;50:1810–1815.
60 Newman AB, Yanez D, Harris T, Duxbury A, Enright PL, Fried LP: Weight change in old age and its association with mortality. J Am Geriatr Soc 2001;49:1309–1318.
61 Wannamethee SG, Shaper AG, Lennon L: Reasons for intentional weight loss, unintentional weight loss, and mortality in older men. Arch Intern Med 2005;165:1035–1040.
62 Gregg EW, Gerzoff RB, Thompson TJ, Williamson DF: Intentional weight loss and death in overweight and obese U.S. adults 35 years of age and older. Ann Intern Med 2003;138:383–389.
63 Fontaine KR, Allison DB: Does intentional weight loss affect mortality rate? Eat Behav 2001;2:87–95.
64 Wing RR, Hamman RF, Bray GA, Delahanty L, Edelstein SL, Hill JO, Horton E S, Hoskin MA, Kriska A, Lachin J, Mayer-Davis E J, Pi-Sunyer, X, Regensteiner JG, Venditti B, Wylie-Rosett J: Achieving weight and activity goals among diabetes prevention program lifestyle participants. Obes Res 2004;12:1426–1434.
65 Garrow JS, Summerbell CD: Meta-analysis: effect of exercise, with or without dieting, on the body composition of overweight subjects. Eur J Clin Nutr 1995;49:1–10.
66 Pavlou KN, Steffee WP, Lerman RH, Burrows BA: Effects of dieting and exercise on lean body mass, oxygen uptake, and strength. Med Sci Sports Exerc 1985;17:466–471.
67 Binder EF, Schechtman KB, Ehsani AA, Steger-May K, Brown M, Sinacore DR, Yarasheski K E, Holloszy JO: Effects of exercise training on frailty in community-dwelling older adults: results of a randomized, controlled trial. J Am Geriatr Soc 2002;50:1921–1928.
68 Seguin R, Nelson ME: The benefits of strength training for older adults. Am J Prev Med 2003; 25(3 suppl 2):141–149.
69 Chang JT, Morton SC, Rubenstein LZ, Mojica WA, Maglione M, Suttorp MJ, Roth EA, Shekelle PG: Interventions for the prevention of falls in older adults: systematic review and meta-analysis of randomised clinical trials. BMJ 2004;328:680.
70 Miller ME, Kral JG: Surgery for obesity in older women. Menopause Int 2008;14:155–162.
71 St Peter SD, Craft RO, Tiede JL, Swain JM: Impact of advanced age on weight loss and health benefits after laparoscopic gastric bypass. Arch Surg 2005; 140:165–168.

72 Sosa JL, Pombo H, Pallavicini H, Ruiz-Rodriguez M: Laparoscopic gastric bypass beyond age 60. Obes Surg 2004;14:1398–1401.
73 Sugerman HJ, DeMaria EJ, Kellum JM, Sugerman EL, Meador JG, Wolfe LG: Effects of bariatric surgery in older patients. Ann Surg 2004;240:243–247.
74 Busetto L, Angrisani L, Basso N, Favretti F, Furbetta F, Lorenzo M: Safety and efficacy of laparoscopic adjustable gastric banding in the elderly. Obesity (Silver Spring) 2008;16:334–338.

Professor Ian Chapman
Division of Medicine University of Adelaide, Level 6 Eleanor Harrald Building
Royal Adelaide Hospital, North Terrace
Adelaide 5000 (Australia)
Tel. +61 8 82224162, Fax +61 8 82233870, E-Mail ian.chapman@adelaide.edu.au

Mobbs CV, Hof PR (eds): Body Composition and Aging.
Interdiscipl Top Gerontol. Basel, Karger, 2010, vol 37, pp 37–50

Central Control of Food Intake in Aging

Zbigniew Kmiec

Department of Histology, Medical University of Gdansk, Gdansk, Poland

Abstract

Energy homeostasis and fuel metabolism undergo significant modifications in the course of aging. This presents in elderly subjects either as increased body mass and glucose intolerance – which may lead to obesity and type 2 diabetes – or loss of appetite, which may also seriously compromise health. The hypothalamic expression of neuropeptide Y (NPY), the most potent orexigen, and its receptors, was highly suppressed in old rats. Moreover, induction of the NPY-dependent responses was severely blunted in old animals. Similar reductions, although of a lower magnitude, were reported for other hypothalamic orexigens, AgRP and orexins. Orexigenic activity of ghrelin, the only peripheral orexigen, was clearly suppressed in old humans and rats. However, aging did not alter hypothalamic expression of key anorexigens, α-MSH and CART. Age-related decrease of central anorexigenic action of leptin was likely caused by the impaired leptin signal transduction. Thus, aging in rodents is associated with the general down-regulation of orexigenic hypothalamic peptides – and unchanged expression of anorexigenic hypothalamic peptides – which may lead to weight loss at the end of life. If similar changes at the level of CNS underlie the 'anorexia of aging' observed in some elderly, therapeutic interventions at this regulatory level may be possible in the future.

Aging in humans and other mammals is associated with failure to maintain homeostasis in response to physiological and environmental challenges. Disturbances of fuel and energy homeostasis occur during aging in three major forms: glucose intolerance and type 2 diabetes, increased accumulation of fat leading to obesity, and, paradoxically, anorexia of aging resulting in the decrease of body mass or even cachexia. In all these states there is a clear imbalance between intake of energy equivalents and their usage. It is hoped that understanding of the age-associated alterations in the regulation of food intake may lead to the development of potential therapeutic approaches. These approaches could counteract the disintegration of metabolic homeostasis that results in obesity, diabetes, or loss of body mass. This review aims at the presentation of recent developments in the field of central regulation of food intake during aging.

Main Components of Central Regulatory Mechanisms of Food Intake

It was proposed half a century ago that the regulation of food intake is based on the activities of two functional hypothalamic centers. On the basis of electrophysiological observations and lesions of various hypothalamic structures, the 'hunger center' – responsible for the activation of food intake and feeding behavior – was localized to the lateral hypothalamic area (LHA) or nucleus. Because lesions of the ventromedial hypothalamic nucleus (VMN) were associated with hyperphagia and abnormal weight gain, this hypothalamic area was called the 'satiety center'. The VMN was shown to be activated by the nervous, hormonal, and metabolic inputs from the gastrointestinal tract after commencing food consumption. This elegant hypothesis was undermined during the last 20 years by the discovery of numerous neuropeptides and their receptors involved in the control of food intake. The distribution of these peptides throughout the CNS was determined, and the electrophysiological responses of respective brain areas to factors known to influence food intake and feeding behavior were characterized [1–4]. These brain areas include the nucleus of the solitary tract (NTS) in the caudal brainstem, where integration of peripheral satiety signals with hypothalamic and suprahypothalamic information takes place. In recent years, it became clear that, in addition to the VMN and LHA, other hypothalamic nuclei were involved in appetite control. It was also shown that hypothalamic neurons express receptors not only for neuropeptides and neurotransmitters, but also for many hormones and metabolites released from peripheral tissues. The discovery of leptin and its synthesis in the adipose tissue [5] was a milestone in the elucidation of pathways used by peripheral hormones and metabolic signals that affect short- and long-term control of energy homeostasis. Currently, more than 50 factors are known to act at the CNS level to stimulate or inhibit food intake [6–9]. Data presented in table 1 show that there is a redundancy of peripheral signals that inhibit ingestive behavior, while only one peripheral hormone, ghrelin, stimulates food intake [10].

It has been well-established that continuous short-time crosstalk between the hypothalamic NPY network and the hormone leptin is essential for energy homeostasis [1–10]. These interactions take place predominantly at the level of two types of ARC (hypothalamic arcuate nucleus) neurons. Orexigenic 'first order' neurons co-express NPY and AgRP (Agouti-related protein), whereas anorexigenic 'first order' neurons co-express the key anorexigenic peptide, melanocortin (α-MSH, α-melanocyte-stimulating hormone), and its precursor propeptide, POMC (proopiomelanocortin). The majority of the POMC/α-MSH neurons express another anorexigenic peptide, CART (Cocaine and amphetamine-regulated transcript). The main area of NPY production is the ARC, and the release of NPY and co-expressed neurotransmitters in other hypothalamic areas – including the PVN (paraventricular nucleus), VMH and LHA – evokes appetite. Moreover, since AgRP is an endogenous inverse agonist to the melanocortin receptors, the release of NPY and AgRP in the ARC-PVN axis restrains

Table 1. Neuropeptides and hormones that control food intake

Centrally-released factors that stimulate food intake	CNS factors that inhibit food intake
AgRP	α-MSH (POMC derivative)
NPY	CART
Orexin A and B	CRH and urocortins
MCH (melanin-concentrating hormone)	TRH
Galanin	GALP (galanin-like peptide)
GHRH	oxytocin
B-endorphin	GLP-1 (glucagon-like peptide 1)
Dynorphin	serotonin
GABA	prolactin
Noradrenalin	dopamine (acting in ARC and LHA)
Dopamine (acting in VMH)	neurotensin and neuromedins
Endogenous cannabinoids	nesfatin-1
Peripheral orexigen	**Peripheral anorexigenic signals**
Ghrelin	leptin, adiponectin
	insulin and amylin
	enterostatin
	bombesin and GRP (gastrin-releasing peptide)
	oxyntomodulin
	CCK (cholecystokinin), $PYY_{3\text{-}36}$ (peptide YY)
	GLP-1, GLP-2, GIP (gastric inhibitory peptide)
	estrogens
	cytokines (IL-1, IL-6, IL-8, TNF, IFN)

Summary of data reviewed by Kalra et al. [1], Woods and A'Allessio [6], Valassi et al. [8] and Sánchez-Lasheras et al. [9].

anorexic melanocortin signaling, thus counteracting the anorexic actions of α-MSH. The NPY/AgRP and POMC/α-MSH neurons not only project to other hypothalamic and CNS nuclei involved in the control of food intake and feeding behavior, but also reciprocally inhibit their own activities depending on the central or peripheral signals which interact with their respective receptors (fig. 1).

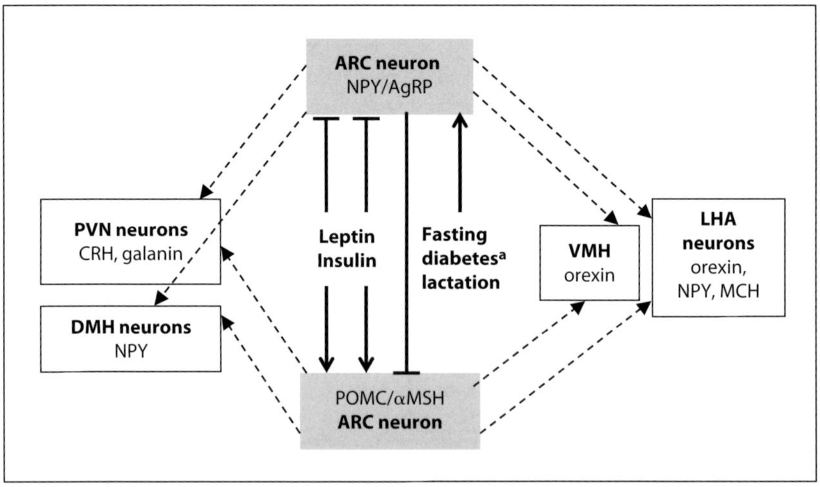

Fig. 1. The interactions between main hypothalamic nuclei involved in the regulation of food intake. Only key neuropeptides have been listed. Lines show major interneuronal connections. The stimulatory and inhibitory effects of peripheral hormones and (patho)physiological states on hypothalamic neurons are shown, respectively, by solid lines ended with arrows or rectangle bars. [a] Interactions between hyperglycemia and prolonged central leptin insufficiency such as observed in aging has been discussed by Kalra [57], recently.

Leptin, insulin, glucose, and cholecystokinin (CCK) belong to the major peripheral signals released in the body in response to food ingestion, while ghrelin levels increase in response to fasting. These agents are transported into the hypothalamus through the blood-brain barrier, which is less tight in the ARC than in other brain areas [11]. Leptin and insulin inhibit NPY expression by binding to their receptors on NPY/AgRP neurons, while simultaneously stimulating POMC/α-MSH neurons in the ARC, resulting in the termination of food intake. Conversely, ghrelin activates NPY/AgRP neurons, thus stimulating orexigenic behavior (fig. 1). However, it has to be mentioned that this very simplistic description of the major regulatory pathways which control food intake, and ultimately body mass, does not include the variety of interrelated neurocircuits involved in the regulation of one of the most complicated and most important aspect of body homeostasis.

Food Consumption in Aging Humans and Laboratory Animals

There is an overwhelming agreement that during normal human aging, a gradual decline in appetite and food consumption – termed the 'anorexia of aging' [12] – takes place.

The age-related decrease of food intake, observed also in laboratory animals, has been regarded as a normal reaction to the diminished energy expenditure that

Table 2. Changes of body weight and food consumption during the life span of some rat strains

Rat strain, sex	Age months	Change with age		Ref. No.
		body weight	food intake g/100 g body weight	
Wistar, male	3 vs. 24	moderate increase	increase	14
Wistar, male	5 vs. 24	moderate increase	no change	23
Wistar, male	4 vs. 26	moderate increase	decrease by 40%[1]	24
Fischer 344, male	3 vs. 20	high increase	decrease	18
Brown Norway, male	4 vs. 20	moderate increase	decrease	22
Long Evans	3, 12, 24, 27	decrease after 24 months	decrease after 24 months	33

[1] Decrease of water intake by 20%.

characterizes healthy elderly. Another feature of normal aging is the gradual decrease of muscle mass (lean body mass) and an increase of fat content. Thus, diminished food intake by the elderly may represent a physiological adaptation to decreased physical activity. However, since aging is associated with important alterations of food perception, many elderly do not consume enough food, which leads to the decline of total and lean body mass, a dangerous health risk factor. Although the age-related decline of appetite has been known for many years, there is little understanding of its physiological basis, especially with respect to the central regulation of ingestive behavior.

Because experimental verification of the central mechanisms of disturbed food intake cannot be performed in humans, answers to this problem were sought in animal experiments in which aged rats represented the most commonly used species. However, in rodents, as in humans, the genetic interindividual and interstrain differences seem to have a big influence on the control of energy homeostasis and body mass during aging. Generally, one can distinguish two types of rat strains with respect to the changes of body weight during their life span. In the first type – represented by Wistar and Brown Norway (BN) rats – animals only modestly increase their weight after reaching maturity, whereas in the second type – Sprague-Dawley (SD) and Fischer 344 (F344) – rats increase their mass more or less constantly throughout their life. Some rat or mouse strains (Zucker rat, *db/db* and *ob/ob* mice) that were extensively used in the studies of energy homeostasis cannot be regarded as valid models for aging research since they show a mutated phenotype, which severely affects the aging process. Examples of the differences in food consumption during aging of various rat strains are shown in table 2.

Age-Related Differences in the Expression and Action of Neuropeptides and Hormones Involved in the Control of Food Intake

NPY

Neuropeptide Y is the best characterized, and probably the most important, orexigenic peptide [1–4, 6–9]. The effect of aging on the gene expression of NPY in fed and fasted rats was therefore evaluated by several authors. Using in situ hybridization, it was found that in the hypothalamus of old Wistar [13, 14] and BN rats, the NPY mRNA levels were several times lower than in young animals [15, 16]. In old F344xBN rats, however, the hypothalamic content of NPY mRNA did not change with aging [17].

Intracerebroventricular (ICV) administration of neuropeptides and hormones is often used to evaluate their effects on food intake despite the use of supraphysiological doses of these agents [1]. After ICV injection of various doses of NPY, virtually no feeding response was observed in senescent rats, in contrast to the dose-dependent high increase of food intake in young or middle-aged animals [18–21].

Alterations in the nutritional state, whether short-term or chronic, profoundly affect various aspects of metabolic and hormonal homeostasis. Fasting and subsequent refeeding belong to nutritional manipulations which reflect a naturally occurring periodicity in nutrients' supply. Many authors reported that with fasting, young rats lost more body weight than older rats, but then regained body mass more rapidly during refeeding [22, 23]. Fasting, which causes a substantial decline of blood leptin and insulin concentrations both in young and old animals [20, 23, 24], was shown to prominently increase hypothalamic NPY expression only in young animals, since in old rats this adaptive response was strongly suppressed. In young BN rats, fasting strongly enhanced hypothalamic mRNA levels of NPY [15, 20, 22, 25] and AgRP [15], and moderately decreased POMC mRNA level [20], while these responses were suppressed in old animals.

The process of aging is often accompanied by decreased cell and tissue reactivity to exogenous agents like drugs or hormones. This may be caused by changes in the number or structure of receptors or decreased activity in signaling pathways. By using quantitative PCR, Coppola et al. [26] found a significant decline in the mRNA of the Y1 and Y2 NPY receptors in the paraventricular nucleus of old F344 rats, compared to young ones. Paradoxically, the number of neurons that showed Y1 receptors was twice as high in old animals, whereas the total number of all PVN neurons did not change with age [26]. This age-related increase of NPY type 1 receptors may compensate for the decreased action of this strong orexigenic neuropeptide.

AgRP

Although many hypothalamic nuclei contain NPY/AgRP neurons and both peptides are potent orexigens, their activities are differently affected by aging. Similarly to NPY, the hypothalamic expression of AgRP was suppressed in old BN rats [15].

However, in contrast to the unchanged food intake after an NPY challenge, increase of food intake after ICV administration of AgRP was well preserved in senescent rats in spite of increased serum leptin levels [15]. These findings suggest that age-related decline in the reactivity of hypothalamic neurons to endogenous orexigenic signals (like fasting), or centrally-administered exogenous agents, occurs in specific ways, which probably depend on the type of signal and neuronal circuits involved.

Ghrelin

Ghrelin, a 28-amino-acid peptide synthesized primarily in the stomach [27] – and in smaller amounts in other organs, including some neurons in the dorsal portion of the ARC and outside of the hypothalamus [10, 27] – was discovered as an endogenous ligand of growth hormone secretagogue receptor (GHS-R1). Ghrelin stimulates growth hormone (GH) release and food intake when administered intravenously to rodents and humans or by a ICV route in animals. Ghrelin stimulates food intake primarily by direct activation of NPY/AgRP neurons in the ARC [10]. Since ghrelin is the only orexigenic peripheral agent that counteracts the satiety signal of leptin, its contribution to the control of food intake in the elderly was investigated soon after its discovery. Serum ghrelin concentrations were lower in old rats than in young ones [28] and did not change with age in C57BL/6J male mice [29]. Although some authors reported a decline in serum ghrelin levels in old humans [30, 31], a recent study did not confirm these findings [32]. However, the number of the investigated elderly subjects in all studies was too small to allow for definite conclusions. Ghrelin secretion depends also on the composition of the diet [10], which was different in the studied populations.

Plasma ghrelin levels are regulated by acute feeding states. They rise during fasting and are rapidly suppressed after feeding. Apparently, aging is accompanied by disturbed regulation of the secretory and sensory function of ghrelin-producing neuroendocrine 'A-X'-like cells of the oxyntic glands of the stomach. This is evidenced by a blunted response of serum ghrelin concentration to 3-day fasting in old BN rats, whereas starvation nearly doubled ghrelin level in young animals [28]. Additionally, the reaction of ghrelin-secreting cells to a carbohydrate-rich test meal was impaired in old humans, since the hormone levels did not change postprandially, whereas the decline was substantial in middle-aged individuals [32]. The orexigenic effect of ghrelin becomes significantly suppressed during aging. The increase of food intake after ICV ghrelin administration was significantly lower in old Wistar rats compared with young and middle-aged animals [21]. Similarly, two hours after intravenous ghrelin administration, food consumption in 27-month old Long Evans rats was by ca. 30% lower than in 3-, 12-, and 24-month old animals [33].

Although GH responses to ghrelin administration in the elderly were lower than those seen in young people [34], no age-related effect on ghrelin-induced magnitude and timing of GH secretion was seen in the Long Evans rat [33]. The long-term administration of rat ghrelin to aged mice improved their lean body weight and reduced adiposity [35]. The potency of ghrelin mimetics – which increase GH secretion, and, consequently, IGF-1 serum concentration – has been used recently as a therapeutic modality to stimulate anabolic processes in the elderly [36]. It may thus be concluded that decreased serum ghrelin levels and inadequate orexigenic response to ghrelin may contribute to the age-related impairment of food intake, and, possibly, the anorexia of aging.

Orexins

Orexin A (OX-A) and orexin B (OX-B) are neuropeptides expressed almost exclusively in the lateral hypothalamic area and the spinal cord, areas that have important roles in energy balance through cortical arousal. OX-A is produced in the caudal lateral, posterior, perifornical, and dorsomedial hypothalamic areas. In the lateral hypothalamic area (LHA), OX-A stimulates feeding and activates several feeding-regulatory brain areas. Both OX-A and OX-B strongly induce food intake and state of vigilance [1, 2]. In a rare case of neuroendocrine studies in aged primates, Downs et al. found that in the brains of rhesus macaques ranging in age from <1 year to 32 years, orexin-immunoreactive neurons, localized predominantly in the LHA, showed identical distribution patterns irrespective of age [37]. Similarly, orexin neuron number and serum orexin B concentrations did not differ with age. In contrast, orexin B-immunoreactive axon density in the locus coeruleus, the brain stem region involved in the regulation of sleep and wakefulness, was significantly lower in old animals than that observed in the young or adult animals [37]. In rats, the number of OX-A-positive and OX-B-positive neurons was significantly lower in the LHA in old *versus* young animals [28, 38]. However, the functional consequences of this decline were not assessed. In other studies, ICV administration of OX-A did not stimulate food intake by old rats compared to young animals [21, 39]. When orexin was injected directly into the lateral hypothalamic area of young, middle-aged and old F344 rats, delayed and diminished feeding responses in old macacques were observed [39]. Moreover, in contrast to younger animals who showed elevated c-Fos-ir response to OX-A (Fos-ir, immunoreactivity of c-Fos, is a marker of neuronal activation), old rats exhibited attenuated or absent c-Fos-ir in all brain regions including the hypothalamic paraventricular nucleus and the rostral nucleus of the solitary tract [39], major brain centers involved in the feeding-stimulating function of orexins [1, 2]. Lack of OX-A action might be caused by an age-related decrease of orexin receptors, since a lower number of type 1 orexin receptors (OX-IR) was found in the hypothalamus of old rats [40], and decreased type 2 orexin receptor mRNA levels were observed in the hypothalamus (also in the brain cortex and basal forebrain) of old mice [41]. These findings implicate the OX-A network as another feeding-controlling system affected by aging.

POMC and MSH

In sharp contrast to the age-related decreased expression of the most potent orexigen, NPY, the hypothalamic expression of POMC, the precursor of a potent anorexigenic α-MSH, was similar in young and old rats [15]. However, similarly to other situations which reveal aging-related impairments after environmental challenges, fasting-induced decrease of POMC mRNA was less prominent in the paraventricular nucleus of old as compared to young rats [15].

CART

Although CART, an anorexigenic peptide, is mostly co-localized with POMC in the same neurons, the effect of aging on its expression is different. The expression of CART mRNA was higher in middle-aged and old rats compared to young ones in the arcuate nucleus. It was, however, lower in the paraventricular nucleus [15, 16, 22].

Leptin and Central Leptin Resistance during Aging

The paramount importance of leptin in the long-term regulation of body weight has been clearly demonstrated. Therefore, the effect of aging on serum leptin concentrations, the presence of its receptors in the hypothalamus and other brain structures, and signaling in the CNS, were investigated by many authors. In various strains of aging rodents, either increased (F344xBN [12] and Wistar [17, 42] rats), unchanged (F344 [19] and Wistar [23, 43, 44] rats) or decreased (Wistar [19] rats) serum leptin concentrations were reported, and the variations seemed to be related to genetic differences between rat strains, as discussed by Kmiec et al. [23].

After a decade of research, it became evident that aging is associated with central (and peripheral) leptin resistance. This age-related defect of leptin action at the level of the hypothalamus was demonstrated by the use of various experimental techniques. For example, hyperleptinemia induced by subcutaneous leptin infusion for 7 days reduced food consumption and hypothalamic NPY levels by 50% in young SDxBN rats. In old animals, however, food consumption was reduced by only 20% and hypothalamic NPY concentration was unaffected [45]. In a similar experimental set-up, the resistance of peripheral tissues to leptin action was demonstrated in old SD rats [46].

The mechanism of central leptin resistance during aging may be caused by several factors. One possibility that has not been investigated in the aging studies could be decreased age-related local production of leptin, since leptin mRNA levels were enriched in the rat arcuate nucleus, and expression was regulated by fasting and refeeding [47]. Another cause of the age-related central leptin resistance may be impaired age-related transport of leptin across the blood–brain barrier, since it is widely believed that leptin may cross the blood-brain-barrier by binding to the longest variant of leptin receptors , LRb (formerly denoted as Ob-Rb) in a vessel's wall using a partially saturable transport system [11]. Indeed, a decreased leptin uptake in hypothalamic nuclei, which was partially caused by the diminished hypothalamic expression of the Ob-Rb receptors, was demonstrated in old Wistar rats [48]. However, the same group showed that adiposity plays a key role in the development of aging-associated leptin resistance, since food restriction recovered leptin responsiveness and increased expression of leptin receptors in the hypothalamus of old rats [49].

Aging-associated suppression of leptin action at the CNS level may be also caused by alterations at the level of the signal transduction pathway(s) in the hypothalamic neurons. Leptin binding to LRb, which belongs to class I cytokine receptors, triggers the autophosphorylation of JAK2, which allows the recruitment and phosphorylation of the cytosolic transcription factor STAT3 (signal transducer and activator of transcription protein 3). After translocation to the nucleus, P-STAT3 binds to specific DNA elements to modulate target gene transcription [50]. When leptin was administered intraventricularly to F344xBN young and old rats, reduced sensitivity for STAT3 phosphorylation, as well as diminished leptin-induced P-STAT3 transcription factor binding, was observed in old animals. There were no changes in the total content of STAT3 protein. However, the amount of leptin receptor protein was 50% lower in old rats [51,52]. The authors concluded that impaired leptin signal transduction may be due to either the elevated obesity with age, or due to age itself, or both.

The aging-related impairment in the hypothalamic leptin action may be also related to the expression of a feedback inhibitor of the leptin signaling pathway, SOCS-3 protein. SOCS-3 is a member of the family of the suppressor-of-cytokine-signaling proteins, which inhibit JAK activity and in this way modulate leptin signaling [50]. Peralta et al. showed that the expression of SOCS-3 in the hypothalamus was increased in old, leptin-resistant Wistar rats [53]. Interestingly, moderate food restriction for 3 months partially reverted the increases in SOCS-3 mRNA levels associated with aging [53]. These results suggest that SOCS-3 may be a mediator of hypothalamic leptin resistance in the aged rat and that the hyperleptinemia associated with aging is, at least in part, responsible for the increase of SOCS-3 expression in the hypothalamus.

Insulin Action on Hypothalamic Centers Controlling Feeding Behavior Is Decreased in Aging

Thirty years ago, Woods et al. [54] demonstrated that chronic ICV infusion of insulin reduced the food intake and body weight of baboons. Since then, the expression of insulin receptors and the different elements of insulin signal transduction pathway in hypothalamic nuclei, the activation of insulin signaling in hypothalamic neurons, and the effects of central administration of insulin on food intake and body weight in different animal models, clearly support insulin's role in the regulation of fuel homeostasis also at the central level [6]. Insulin shares some of the hypothalamic targets of leptin. However, the effects of leptin at the CNS level on food intake and energy balance seem to be more efficient than those of insulin. Nevertheless, the data obtained in mice with a neuron-specific disruption of the insulin receptor gene, which present diet-sensitive obesity and mild insulin resistance, among other alterations, clearly demonstrate the importance of insulin as an energy balance regulator, acting through the CNS [55]. García-San Frutos et al. [56] showed that after ICV insulin administration, insulin signaling was clearly impaired in old Wistar rats, both at the level of insulin receptor and IRS-2 (insulin-related substrate) as well as downstream of the PI3K (phosphoinositide-3-kinase) pathway. Interestingly, food restriction improved both insulin responsiveness and insulin signaling [56]. These data suggest that Wistar rats develop hypothalamic insulin resistance with aging and that this is probably caused by the alterations of the signal transduction pathway. This central insulin resistance is accompanied by the age-dependent decreased insulin sensitivity of peripheral tissues (with unchanged serum insulin concentrations in young, adult, and old rats) that has been related mainly to the increased content of retroperitoneal fat in old animals [42].

In addition to aging, leptin resistance accompanies obesity and in most cases insulin resistance. In non-obese animals, both insulin and leptin act on the hypothalamus to inhibit feeding behavior. If the anorexic action of leptin is dependent on normal insulin signaling, insulin resistance would also present as leptin resistance. To test this hypothesis, Matsumoto et al. [25] chronically administered the insulin sensitizer troglitazone (a peroxisome proliferator-activated receptor -α agonist) to old BN male rats. Troglitazone reduced their high insulin, high leptin, and high body fat. Compared to untreated old animals, troglitazone-treated old rats refed after the 72-h fast, had greater food intake, regained more weight at a faster rate, and returned to baseline body weight by the end of the refeeding period. Since troglitazone administration did not alter NPY gene expression in the arcuate nucleus, the restoration of this metabolic phenotype characteristic for young animals might be due to alterations of other neuropetides that regulate energy balance [25]. These results provide an important link between insulin and leptin resistance that apparently contributes to the impairments in energy and weight regulation. However, additional studies are necessary to establish the mechanisms of the interactions between insulin resistance, leptin resistance, and aging.

In conclusion, in rodents, aging is associated with the downregulation of orexigenic hypothalamic peptides with no change in hypothalamic anorexic peptides. Therefore, in the aged rat the delicate balance between hypothalamic orexigens and anorexigens that controls food intake seems to be shifted to the latter. This situation may be responsible for the decreased appetite drive in senescent animals and some elderly, and for loss of weight at their end-of-life period. If the 'anorexia of aging' observed in humans is related to similar alterations in the central control of food intake as found in old animals, the therapeutic intervention at this regulatory level would be possible in the future.

Acknowledgement

The work was supported by grant ST-12 from the Medical University of Gdansk, Poland.

References

1 Kalra SP, Dube MG, Pu S, Xu B, Horvath TL, Kalra PS: Interacting appetite-regulating pathways in the hypothalamic regulation of body weight. Endocr Rev 1999;20:68–100.
2 Leibowitz SF, Wortley KE: Hypothalamic control of energy balance: different peptides, different functions. Peptides 2004;25:473–504.
3 Morton GJ, Cummings DE, Baskin DG, Barsh GS, Schwartz MW: Central nervous system control of food intake and body weight. Nature 2006;443:289–295.
4 Schwartz MW: Central nervous system regulation of food intake. Obesity (Silver Spring) 2006;14:1–8.
5 Zhang Y, Proenca R, Maffei M, Barone M, Leopold L, Friedman JM: Positional cloning of the mouse obese gene and its human homologue. Nature 1994; 372:425–432.
6 Woods SC, D'Alessio DA: Central control of body weight and appetite. J Clin Endocrinol Metab 2008; 93:37–50.
7 Beckers S, Zegers D, Van Gaal LF, Van Hul W: The role of the leptin-melanocortin signalling pathway in the control of food intake. Crit Rev Eukaryot Gene Expr 2009;19:267–287.
8 Valassi E, Scacchi M, Cavagnini F: Neuroendocrine control of food intake. Nutr Metab Cardiovasc Dis 2008;18:158–168.
9 Sánchez-Lasheras C, Christine Könner AC, Jens C. Brüning JC: Integrative neurobiology of energy homeostasis-neurocircuits, signals and mediators. Front Neuroendocrinol 2010;31:4–15.
10 Ferrini F, Salio C, Lossi L, Merighi A: Ghrelin in central neurons. Curr Neuropharmacol 2009;7:37–49.
11 Kastin AJ, Pan W: Dynamic regulation of leptin entry into brain by the blood-brain-barrier. Regulat Pept 2000;92:37–43.
12 Morley JE: Anorexia and weight loss in older persons. J Gerontol A Biol Sci Med Sci 2003;58:131–137.
13 Peralta S, Carrascosa JM, Gallardo N, Ros M, Arribas C: Ageing increases SOCS-3 expression in rat hypothalamus: effects of food restriction. Biochem Biophys Res Commun 2002;296:425–428.
14 Kaneda T, Makino S, Nishiyama M, Asaba K, Hashimoto K: Differential neuropeptide responses to starvation with ageing. J Neuroendocrinol 2001; 13:1066–1075.
15 Wolden-Hanson T, Marck BT, Matsumoto AM: Blunted hypothalamic neuropeptide Y gene expression in response to fasting, but preservation of feeding responses to AgRP in aging male Brown Norway rats. Am J Physiol Regul Integr Comp Physiol 2004; 287:138–146.
16 Sohn EH, Wolden-Hanson T, Matsumoto AM: Testosterone (T)-induced changes in arcuate nucleus cocaine-amphetamine-regulated transcript and NPY mRNA are attenuated in old compared to young male Brown Norway rats: contribution of T to age-related changes in cocaine amphetamine-regulated transcript and NPY gene expression. Endocrinology 2002;143:954–963.
17 Kalra SP, Bagnasco M, Otukonyoung EE, Dube MG, Kalra PS: Rhythmic, reciprocal ghrelin and leptin signaling: new insight in the development of obesity. Regul Pept 2003;111:1–11.
18 Blanton CA, Horwitz BA, Blevins JE, Hamilton JS, Hernandez EJ, Mc Donald RB: Reduced feeding response to neuropeptide Y in senescent Fischer 344 rats. Am J Physiol Regul Integr Comp Physiol 2001;280:1052–1060.
19 Blanton CA, Horwitz BA, Murtagh-Mark C, Gietzen DW, Griffey SM, McDonald RB: Meal patterns associated with the age-related decline in food intake in the Fischer 344 rat. Am J Physiol 1998;275:1494–1502.
20 Wolden-Hanson T: Mechanisms of the anorexia of aging in the Brown Norway rat. Physiol Behav 2006; 88:267–276.
21 Akimoto-Takano S, Sakurai C, Kanai S, Hosoya H, Ohta M, Miyasaka K: Differences in the appetite-stimulating effect of orexin, neuropeptide Y and ghrelin among young, adult and old rats. Neuroendocrinology 2005;82:256–263.
22 Gruenewald DA, Marck BT, Matsumoto AM: Fasting-induced increases in food intake and neuropeptide Y gene expression are attenuated in aging male Brown Norway rats. Endocrinology 1996;137: 4460–4467.
23 Kmiec Z, Pokrywka L, Kotlarz G, Kubasik J, Szutowicz A, Mysliwski A: The effects of fasting and refeeding on serum leptin, adiponectin and free fatty acids concentrations in young and old male rats. Gerontology 2005;51:357–362.

24 Pu S, Dube MG, Kalra PS, Kalra SP: Regulation of leptin secretion: effects of aging on daily patterns of serum leptin and food consumption. Regul Pept 2000;92:107–111.
25 Kotz, CM, Mullett MA, Wang CF: Diminished feeding responsiveness to orexin A (hypocretin 1) in aged rats is accompanied by decreased neuronal activation. Am J Physiol Regul Integr Comp Physiol 2005;289:359–366.
26 Coppola JD, Horwitz BA, Hamilton J, McDonald RB: Expression of NPY Yl and Y5 receptors in the hypothalamic paraventricular nucleus of aged Fischer 344 rats. Am J Physiol Regul Integr Comp Physiol 2004;287:69–75.
27 Kojima M, Hosoda H, Date Y, Nakazato M, Matsuo H, Kangawa K: Ghrelin is a growth hormone-releasing acylated peptide from stomach. Nature 1999;402: 656–660.
28 Porkka-Heiskanen T, Alanko L, Kalinchuk A, Heiskanen S, Stenberg D: The effect of age on prepro-orexin gene expression and contents of orexin A and B in the rat brain. Neurobiol Aging 2004;25: 231–238.
29 Ariyasu H, Iwakura H, Yamada G, Nakao K, Kangawa K, Akamizu T: Efficacy of ghrelin as a therapeutic approach for age-related physiological changes. Endocrinology 2008;149:3722–3728.
30 Rigamonti AE, Pincelli AI, Corra B, Viarengo R, Bonomo SM, Galimberti D, Scacchi M, Scarpini E, Cavagnini F, Muller EE: Plasma ghrelin concentrations in elderly subjects: comparison with anorexic and obese patients. J Endocrinol 2002;175:1–5.
31 Sturm K, MacIntosh CG, Parker BA, Wishart J, Horowitz M, Chapman IM: Appetite, food intake, and plasma concentrations of cholecystokinin, ghrelin, and other gastrointestinal hormones in undernourished older women and well-nourished young and older women. J Clin Endocrinol Metab 2003;88:3747–3755.
32 Bauer JM, Haack A, Winning K, Wirth R, Fischer B, Uter W, Erdmann J, Schusdziarra V, Sieber CC: Postprandial response of active ghrelin and prolonged suppression of hunger sensation were observed in the elderly. J Gerontol A Biol Sci Med Sci 2009;doi:10.1093/gerona/glp174.
33 Toshinai K, Mondal MS, Shimbara T, Yamaguchi H, Date Y, Kangawa K, Nakazato M: Ghrelin stimulates growth hormone secretion and food intake in aged rats. Mech Ageing Dev 2007;128:182–186.
34 Broglio F, Benso A, Castiglioni C, Gottero C, Prodam F, Destefanis S, Gauna C, van der Lely AJ, Deghenghi R, Bo M, Arvat E, Ghigo E: The endocrine response to ghrelin as a function of gender in humans in young and elderly subjects. J Clin Endocrinol Metab 2003;88:1537–1542.
35 Ariyasu H, Iwakura H, Yamada G, Nakao K, Kangawa K, Akamizu T: Efficacy of ghrelin as a therapeutic approach for age-related physiological changes. Endocrinology 2008;149:3722–3728.
36 Nass R, Pezzoli SS, Oliveri MC, Patrie JT, Harrell FE Jr, Clasey JL, Heymsfield SB, Bach MA, Vance ML, Thorner MO: Effects of an oral ghrelin mimetic on body composition and clinical outcomes in healthy older adults: a randomized trial. Ann Intern Med 2008;149:601–611.
37 Downs JL, Dunn MR, Borok E, Shanabrough M, Horvath TL, Kohama SG, Urbanski HF: Orexin neuronal changes in the locus coeruleus of the aging rhesus macaque. Neurobiol Aging 2007;28:1286–1295.
38 Sawai N, Ueta Y, Nakazato M, Ozawa H: Developmental and aging change of orexin-A and -B immunoreactive neurons in the male rat hypothalamus. Neurosci Lett 2010;468:51–55.
39 Kotz, CM, Mullett MA, Wang CF: Diminished feeding responsiveness to orexin A (hypocretin 1) in aged rats is accompanied by decreased neuronal activation. Am J Physiol Regul Integr Comp Physiol 2005;289:359–366.
40 Takano S, Kanai S, Hosoya H, Ohta M, Uematsu H, Miyasaka K: Orexin-A does not stimulate food intake in old rats. Am J Physiol Gastrointest Liver Physiol 2004;287:G1182–G1187.
41 Terao A, Apte-Deshpande A, Morairty S, Freund YR, Kilduff TS: Age-related decline in hypocretin (orexin) receptor 2 messenger RNA levels in the mouse brain. Neurosci Lett 2002;332:190–194.
42 Escrivá F, Gavete ML, Fermín Y, Pérez C, Gallardo N, Alvarez C, Andrés A, Ros M, Carrascosa JM: Effect of age and moderate food restriction on insulin sensitivity in Wistar rats: role of adiposity. J Endocrinol 2007;194:131–141.
43 Nogalska A, Swierczynski J: The age-related differences in obese and fatty acid synthase gene expression in white adipose tissue of rat. Biochim Biophys Acta 2001;1533:73–80.
44 Ishii S, Shibasaki T, Murakami T, Shima K, Wakabayashi I: Response of leptin mRNA to 24-h food deprivation and refeeding is influenced by age in rats. Regul Pept 2000;92:45–50.
45 Scarpace, PJ, Matheny M, Moore RL, Tümer N: Impaired leptin responsiveness in aged rats. Diabetes 2000;49:431–435.
46 Gabriely I, Ma XH, Yang XM, Rossetti L, Barzilai N: Leptin resistance during aging is independent of fat mass. Diabetes. 2002;51:1016–1021.
47 Wilkinson M, Brown R, Imrash SA, Ur E: Adipokine gene expression in brain and pituitary gland. Neuroendocrinology 2007;86:191–209.

48 Fernandez-Galaz C, Fernandez-Agullo T, Campoy F, Arribas C, Gallardo N, Andres A, Ros M, Carrascosa JM: Decreased leptin uptake in hypothalamic nuclei with ageing in Wistar rats. J Endocrinol 2001;171:23–32.
49 Fernández-Galaz C, Fernández-Agulló T, Pérez C, Peralta S, Arribas C, Andrés A, Carrascosa JM, Ros M: Long-term food restriction prevents aging associated central leptin resistance in Wistar rats. Diabetologia 2002;45:997–1003.
50 Münzberg H, Björnholm M, Bates SH, Myers MG: Leptin receptor action and mechanisms of leptin resistance. Cell Mol Life Sci 2005;62:642–645.
51 Scarpace PJ, Matheny M, Tümer N: Hypothalamic leptin resistance is associated with impaired leptin signal transduction in aged obese rats. Neuroscience 2001;104:1111–1117.
52 Scarpace PJ, Matheny M, Shek EW: Impaired leptin signal transduction with age related obesity. Neuropharmacology 2000;39:1872–1879.
53 Peralta S, Carrascosa JM, Gallardo N, Ros M, Arribas C: Ageing increases SOCS-3 expression in rat hypothalamus: effects of food restriction. Biochem Biophys Res Commun 2002;296:425–428.
54 Woods SC, Lotter EC, McKay LD, Porte D Jr: Chronic intracerebroventricular infusion of insulin reduces food intake and body weight of baboons. Nature 1979;282:503–505.
55 Könner AC, Klöckener T, Brüning JC: Control of energy homeostasis by insulin and leptin: targeting the arcuate nucleus and beyond. Physiol Behav 2009;97:632–638.
56 García-San Frutos M, Fernández-Agulló T, De Solís AJ, Andrés A, Arribas C, Carrascosa JM, Ros M: Impaired central insulin response in aged Wistar rats: role of adiposity. Endocrinology 2007;148:5238–5247.
57 Kalra SP: Disruption in the leptin–NPY link underlies the pandemic of diabetes and metabolic syndrome: New therapeutic approaches. Nutrition 2008;24:820–826.

Prof. Dr. Zbigniew Kmieć
Department of Histology, Medical University of Gdansk
ul. Debinki 1
80–211 Gdansk (Poland)
Tel. +48 58 3491437, Fax +48 58 3491419, E-Mail zkmiec@amg.gda.pl

Mobbs CV, Hof PR (eds): Body Composition and Aging.
Interdiscipl Top Gerontol. Basel, Karger, 2010, vol 37, pp 51–63

Changes in Food Intake and Its Relationship to Weight Loss during Advanced Age

Roger B. McDonald · Rodney C. Ruhe

Department of Nutrition, University of California, Davis, Calif., USA

Abstract

The results of extensive human and animal studies suggest that declining food intake and body weight observed in the later stages of life may be part of the normal progression of physiological decline observed during aging. Proposed etiologies cover a wide range of biological and psychological conditions. Studies in humans suggest an imbalance in homeostatic mechanisms governing hunger and satiety. That is, while older vs. younger individuals retain a similar drive (hunger) to eat, satiety occurs sooner during a meal in aged people and leads to an overall decrease in daily food intake. Age-related weight loss and a reduction in food intake have also been observed in laboratory animals. Alterations in neurochemical control of energy balance, especially as they relate to long-term regulation of food intake, have received much attention in recent years as the likely mechanism underlying age-related spontaneous weight loss. Age-related changes to neuroendocrine factors such as neuropeptide Y, GABA, CCK, leptin, and insulin have been linked to spontaneous weight loss observed during late life. This brief review provides an update on putative mechanisms underlying the dysregulation of feeding during advanced age that result in body weight loss.

Introduction

Human beings evolved successfully due, in part, to the ability to extract sufficient energy and nutrients from a variety of food sources and under a wide range of environmental conditions. Underlying this evolutionary success is a complex physiological system that controls food intake and energy expenditure, resulting in the maintenance of energy homeostasis. Because of its critical importance to survival and reproduction, food intake is controlled by a system that is tightly regulated and highly redundant, comprising neural and chemical signals that interact in a cooperative, synergistic, and even antagonistic manner to modulate feeding behavior. The complexity and redundancy of the system insures that a disruption in one control mechanism of food intake does not compromise overall energy balance. A complete description of food intake regulation lies well beyond the scope of this review and the reader should consult previous publications [1] for a more thorough description of this process.

In the absence of overt disease, food intake and energy homeostasis are generally well maintained throughout the reproductive period (18–50 years of age) and for many years beyond. However, in the latter part of the 20th century, as lifespan increased and the population of the elderly grew larger, clinicians described a set of symptoms associated with an idiopathic decline in health and well-being of individuals who were approaching the end of life. These symptoms included involuntary weight loss, attenuated food intake, increased fragility, depression, and a general lack of social competence. Collectively, these symptoms were referred to as geriatric failure to thrive because the patients did not respond to interventions and were likely to die within weeks of admission to hospital. Subsequent clinical and experimental investigations into the mechanism underlying geriatric failure to thrive suggested that involuntary weight loss caused by a decline in appetite (food intake) was the initial manifestation of the disorder. So common and pervasive were the presentation of weight loss/attenuated food intake that the phrase 'anorexia of aging' became a more common descriptor for the collection of signs and symptoms associated with geriatric failure to thrive.

The anorexia of aging can occur without an obvious cause in otherwise healthy individuals. Morley et al. [2] were among the first to suggest that a decline in food intake and significant weight loss may be normal physiological responses to advanced age. This suggestion was based largely on clinical observations and the results of longitudinal studies, including the first National Health and Nutrition Examination Survey (NHANES) that described a significant decline in energy intake with age and undernutrition in the advanced age groups [3]. The NHANES data were later confirmed by subsequent NHANES Surveys (fig. 1) [4], the Continuing Survey of Food Intakes by Individuals (fig. 1) [5], and the Baltimore Longitudinal Study [6]. All of these studies demonstrated a progressive decrease in caloric intake in older adults. In addition, short-term dietary intake observations indicate reduced caloric intake in free-living elderly populations [7].

Although the anorexia of aging has become an accepted part of the gerontological vocabulary and is viewed by many as a normal physiological consequence of late life, the mechanism underlying this syndrome remains to be elucidated. A disruption in the normal regulation of energy balance (calorie intake – calorie expenditure), psychological and social variables, and pathological conditions all contribute to the development of this syndrome. The purpose of this review is to summarize current knowledge regarding the regulation of food intake and its relationship to body weight loss during advanced age in the absence of pathological conditions. The results of extensive human and animal studies will be presented to support the position that the reduction in food intake and decline in body weight observed in the later stages of life contribute, in part, to the normal progression of physiological decline observed during aging.

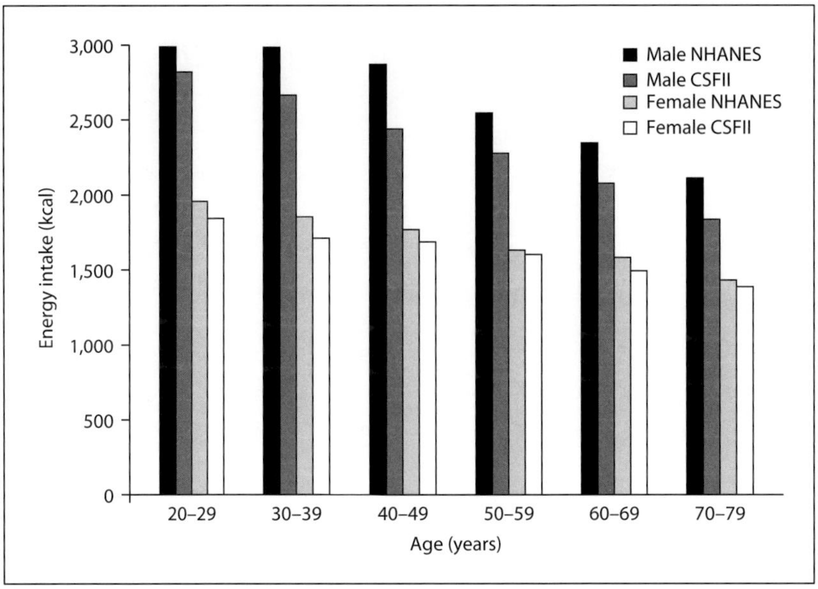

Fig. 1. Daily energy intake from the Third National Health and Nutrition Examination Survey (NHANES; [4]) and the Continuing Survey of Food Intakes by Individuals (CSFII; [5]).

Declining Food Intake during Aging: The Satiety Factor

Age-associated changes in body weight over the life span to do not correlate with decreasing food intake (fig. 2). That is, a general increase in body weight occurs during the third through sixth decades of life while food intake decreases. An age-related decline in body weight does not begin until approximately the seventh decade of life, although food intake during this period declines at approximately the same rate as that observed in the younger group. The decline in body weight seen during the seventh to ninth and greater decades of life cannot be attributed to increased energy expenditure [8]. Thus, one conclusion that can be drawn from this apparent paradox of decreased food intake and increased body weight observed late in the life span is that the 'normal' mechanisms of energy homeostasis are altered during aging.

An imbalance in homeostatic mechanisms governing hunger and satiety, resulting in alterations in energy intake, occur as a normal part of biological aging. The alterations in energy intake regulation are particularly apparent beginning in the sixth to seventh decades of human life. In evaluating the effect of dietary glucose supplementation on gastric emptying and food intake in older (age 65–84 years) men and women, Beckoff et al. [9] showed that subjects did not modify their diets to compensate for additional energy consumed. Rolls et al. [10], using a preloading method, demonstrated that elderly men (60–84 years old) did not compensate as precisely

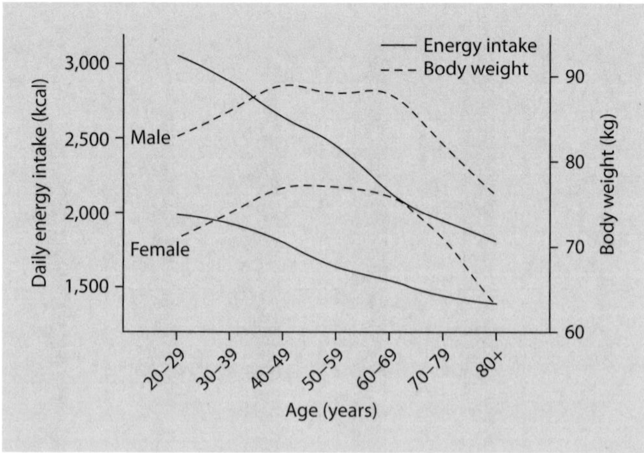

Fig. 2. Daily energy intake and body weight of adults. Data taken from NHANES III.

for changes in caloric intake as did younger men (18–35 years old). In agreement with these data, Roberts et al. [11] found that older men (mean age 68 years) were less successful than younger men (mean age 22.7 years) in adjusting energy intakes after their diet was altered (i.e. elderly subjects did not self-regulate food intake and resume their 'normal' body weights after periods of under- or overfeeding, as did the younger men). In addition, less hunger and greater fullness were reported by the older subjects in this study. Given the short-term nature of both the Roberts et al. [11] and Rolls et al. [10] investigations, the effect of a longer period of underfeeding and weight regain remained unclear. These effects were addressed in an investigation in which underfed younger and older men and women were underfed for 6 weeks followed by 6 months of monitoring their body weight and food intake [12]. Old subjects lost considerably more weight than did younger individuals during the 6 weeks of underfeeding. Moreover, old vs. younger subjects were unable to regain the weight to pre-underfeeding values, a finding with strong implications for the health of older individuals. That is, voluntary weight loss through a reduction in caloric intake during advanced age may trigger involuntary weight loss leading to complications of the anorexia of aging.

The investigations discussed above [10, 11] provide clear evidence that the control mechanisms underlying food intake are altered in aging. The precise nature of these mechanisms remains unclear. Early studies attributed the attenuation in the desire to eat in aged individuals to age-related alterations in olfaction and taste [13]. These studies, however, included persons with frank pathologies, unknown medical histories, and age-related neurological disorders that can attenuate food intake by disrupting normal taste, smell, and swallowing. More recent investigations, in which measures were taken to exclude diseased subjects, have shown that while there appears to be some attenuation of the ability to taste and smell with age, the preponderance of evidence indicates that smell and taste deficits are not contributing factors to the age-related decline in food intake regulation [14].

Gastrointestinal dysfunction has also been considered as a possible inhibitor of normal food intake in elderly humans. Many studies showing reduced gastric secretion of hydrochloric acid and the subsequent inhibition of digestion, absorption, and gastric emptying at older ages did not exclude the presence of atrophic gastritis or its causative bacterium, *Helicobacter pylori*. In the absence of this age-related disease and its causative agent, there appears to be no significant reduction in gastric acid secretion, transit time, or nutrient absorption [15]. Thus, there is currently insufficient evidence to implicate gastrointestinal dysfunction as a primary contributor to weight loss in advanced ages in individuals without atrophic gastritis.

Hypothalamic Correlates of Declining Body Weight during Aging

Neurocenters in the hypothalamus regulate, in part, several basic functions of mammalian physiology including food intake, body weight, temperature, and circadian rhythm. It follows, then, that attenuated hypothalamic function may be associated with age-related alterations in energy balance observed in humans. Only a few correlative studies in humans have been reported that suggest a link between hypothalamic function and age-related decline in food intake and body weight. These reports have described generally that altered temperature regulation and circadian sleep rhythm are often part of the physiopathology associated with the anorexia of aging [16]. Others have found age-related alterations in serum levels of neuroendocrine factors that affect hypothalamic regulation of food intake, such as insulin and leptin [17], findings suggestive of a disruption in the central-mediated regulation of body weight.

Most of the knowledge concerning the relationship among hypothalamic function, energy homeostasis, and aging has come from experiments in laboratory animals. Declining body weight near the end of the natural life span in rodents has been known since the first well-controlled life span studies of McCay et al. [18] during the 1930s. Our laboratory first reported a possible connection between hypothalamic function and age-related body weight loss in rodents while evaluating the effects of aging on temperature regulation [19]. We found significant variation in core temperature between individual old (26–28 months), but not young (6–12 months of age), Fisher rats exposed to a temperature of 6°C for 6 h. While the mean cold-exposed core temperature in old vs. young rats was slightly less, some of the old animals, but none of the young animals, became hypothermic within 2–3 h of the cold exposure. A retrospective analysis of our old rats revealed that the lowest cold-exposed core temperatures were associated with the lowest body weights [unpubl. data].

Results from the thermoregulation investigations led us to hypothesize that age-related differences in cold exposure were more closely related to body weight than to

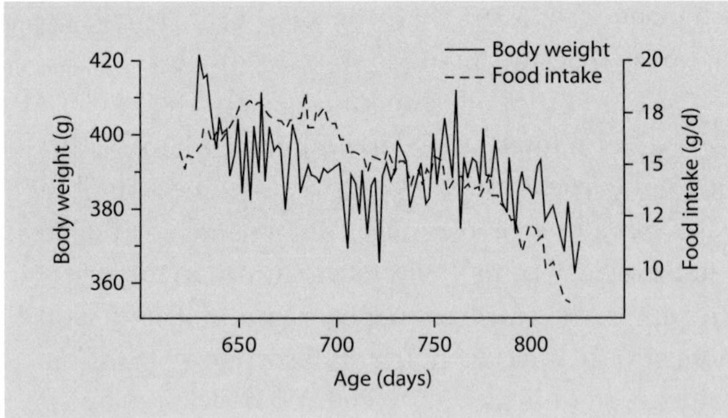

Fig. 3. Relationship between body weight and food intake in a male F344 rat. The rat died at 839 days of age.

age per se. This hypothesis was tested by exposing old and young rats to cold during periods of stable body weight and then exposing the same old rats during a period of declining body weight [20]. We found that 22- to 28-month-old rats with stable body weights were capable of maintaining core temperatures during cold exposure and that their values did not differ significantly from those of younger animals. However, when these same old rats began to lose body weight they were no longer able to thermoregulate in the cold and developed hypothermia. These studies also revealed two additional findings important to the relationship between body weight and longevity in rats: (1) all old animals regardless of age lost body weight near the end of life, and (2) the start of age-related weight loss correlated with the end of the natural life span – death occurred within 3–6 weeks of the start of weight loss.

Oxygen consumption measured during weight-stable and spontaneous weight-loss periods in old rats showed that increased energy expenditure did not explain the spontaneous loss in body weight near the end of life [20]. Therefore, we evaluated food intake in old rats during periods of stable body weight and spontaneous weight loss. As expected, a decline in food intake was highly correlated with the start of weight loss in the old rats (fig. 3). The declining food intake of the old rats could not be explained by altered taste as rapid-weight-loss animals preferred the same flavored food they ate during their weight-stable period [21]. Rather, we found that declining food intake reflected more closely the meal pattern of the old rat. Old spontaneous-weight-loss rats spent less time at the food cup and ate significantly smaller meals then they did during their weight-stable period. The number of meals eaten per day did not differ significantly. These data suggest that simultaneous with rapid spontaneous weight loss, aging rats near the end of their life span demonstrate significant changes in feeding patterns suggestive of earlier satiation, a suggestion consistent with the human studies of Roberts et al. [11] and Rolls et al. [10].

Results from studies of human and laboratory animals strongly indicate that an age-related disruption in food intake control leading to body weight loss reflects a central-mediated factor, most likely involving the hypothalamus. Our laboratory evaluated the overall function of the hypothalamus by measuring circadian rhythm of temperature (CRT), a function controlled almost entirely by the suprachiasmatic nucleus within the hypothalamus [22]. We found that older weight-stable animals, regardless of age, maintained period and amplitude of CRT consistent with patterns observed in younger animals. However, the transition from weight-stable to spontaneous weight loss resulted in a disruption in overall CRT; neither period nor amplitude followed a predictable pattern. The disruption in CRT observed in the spontaneous-weight-loss rats could not be attributed to a single end-of-life pathology or physical changes to the hypothalamic-pituitary interface [23]. Together, the data from our laboratory and those of several others strongly point to the hypothalamus and/or neuroendocrine factors that affect hypothalamic regulation of food intake as playing a pivotal role in the age-related spontaneous weight loss observed in both humans and laboratory animals.

Neuropeptide Y

Hypothalamic neuropeptide Y (NPY) has a powerful stimulatory effect on food intake and energy storage. The concentration of NPY in the hypothalamic paraventricular nucleus (PVN) increases during acute and long-term calorie restriction [24] whereas repeated injections of NPY into the PVN increases food intake and leads to obesity [25]. The potent effect of NPY on enhancing energy storage reflects its role in mechanisms that regulate whole body energy balance. Injection of insulin and/or leptin into the arcuate nucleus (ARC) inhibits the synthesis of NPY that, in turn, prevents hyperphagia (see review [26]). While NPY has been viewed primarily as a long-term regulator of food intake, the antihyperphagic effect of insulin suggests that NPY may also have a role in the regulation of satiety.

Pich et al. [27] reported that food intake in old vs. young, 24-hour-fasted Sprague-Dawley rats was significantly less after injection of NPY into the PVN. While these data clearly indicate attenuation of the stimulatory effect of NPY on food intake in old rats, they do not address the issue of age-related loss in body weight. To this end, we characterized the food intake patterns of young and old weight-stable and spontaneous-weight-loss rats following the injection of NPY into the lateral ventricles of the hypothalamus [23]. Young and old weight-stable rats responded similarly to the food intake stimulatory effects of NPY. However, the same old rats experiencing spontaneous weight loss had significantly smaller increases in the size and duration of their first post-NPY meal. Together, our data and that of Pich et al. [27] imply that the blunted response of food intake to NPY may be a factor in age-related body weight loss.

The attenuated response to NPY observed in the old spontaneous-weight-loss rats could be associated with disruption in NPY synthesis pathways or loss of neural tissue within the PVN. A significant age-related decline in NPY concentration within various rat brain regions has been reported, suggesting altered synthesis rates [28]. The loss of NPY neurons in the cerebral cortex of old (29 month) versus young (4–6 months) Sprague-Dawley rats has been taken to mean that NPY circuitry may be susceptible to age-related declines [28]. However, we were unable to find a significant difference in the number of neurons or number of NPY receptors within PVN isolated from old spontaneous-weight-loss rats compared to those isolated from young animals or old age-matched, weight-stable rats [29]. Although NPY concentration was not measured directly in our studies, it is unlikely that receptor number would remain at young or old weight-stable values during a time of decreasing NPY concentration. Thus, our data suggest that loss of neural tissue and/or decreased concentration of NPY (synthesis) within the PVN cannot account for the attenuation of food intake observed in the spontaneous-weight-loss old rats.

Gamma-Amino Butyric Acid

Gamma-amino butyric acid (GABA) is the brain's primary inhibitory neurotransmitter and is found in high concentration within the lateral hypothalamus, the area most associated with food intake and body weight regulation. GABA's effect on food intake remains controversial, with one investigator finding that chronic activation of the $GABA_A$ receptor decreases food intake [30] while another reported that GABA potentiates NPY-stimulated food intake [31]. We explored the possibility that the decreased food intake observed in spontaneous-weight-loss old rats reflects alteration in the actions of GABA [32]. To this end, various doses of the GABA agonist muscimol, NPY, and NPY + muscimol were injected into the PVN of young, old weight-stable and old spontaneous-weight-loss rats. Although muscimol or NPY administered separately enhanced food intake in the young and old weight-stable rats, it did not do so in the old spontaneous-weight-loss animals. Moreover, and unlike the case in the young and old weight-stable rats, the effect of muscimol plus NPY injection was not greater than that of NPY itself. Thus, the decrease in food intake and body weight observed in the old spontaneous-weight-loss rats reflects attenuated responsiveness to GABA, as well as to NPY. The fact that GABA and NPY administered alone or together reduced food intake seems to suggest a more widespread mechanism than alterations in neuropeptide actions. That is, these data are consistent with the hypothesis that a loss in body weight near the end of life most likely reflects a general decline in hypothalamic function.

Leptin

Leptin is a polypeptide hormone secreted by adipose tissue to inform the brain about adiposity levels of the body. Its actions on body weight regulation have been reviewed recently [33]. The binding of leptin to receptors in the hypothalamus inhibits gene expression of orexigenic peptides NPY and agouti-related peptide and stimulates expression of the anorexic peptides pro-opiomelancortin and cocaine-amphetamine-related transcript. The observation that circulating leptin levels are positively correlated with adiposity suggests that the age-related increase of fat mass seen in humans and laboratory animals may be related to leptin insensitivity.

Leptin's role in body weight loss late in the life span remains largely unknown and we are unaware of data directly evaluating serum leptin concentration in humans experiencing age-related spontaneous weight loss. Studies in healthy aged humans and rodents suggest a similar leptin response to a reduction in body fat as that observed in younger individuals [34, 35]. That is, as body fat/body weight decreases so too does the serum concentration of leptin. In our model of age-related spontaneous weight loss, we find that serum leptin concentration in old rats losing weight declines slightly as compared to older and younger weight-stable animals [23]. Declining serum leptin concentration observed in the old spontaneous-weight-loss rats did not, however, prevent declining food intake as would be predicted from leptin's inhibitory effect on orexigenic neuropeptides. This finding suggests that age-related spontaneous weight loss may reflect a dysfunction in leptin signaling within the hypothalamus. Although previous investigations have demonstrated a decrease in the density of hypothalamic leptin receptors in aged obese rats [36], similar studies have not been conducted in aged rats experiencing spontaneous weight loss.

Insulin

The role of insulin as a long-term modulator of energy homeostasis is well established [1]. Both fasting and postprandial insulin levels are correlated with body adiposity in humans and reviews have recently discussed possible mechanisms [37]. Moreover, insulin receptors are plentiful within the ARC of the hypothalamus [38] and central insulin administration inhibits NPY [39]. Taken together, these findings imply that elevated serum concentrations of insulin could be one mechanism underlying age-related body weight loss.

The impact that insulin may have on age-related body weight loss has not been investigated directly. Rather, alterations in insulin secretion and serum concentration observed in aging humans have been cited as indirect evidence for a possible role in body weight regulation. However, research in this area has yielded mixed results, in part due to the use of subjects of varying age, levels of health, physical activity, and adiposity. Studies of mature, nonobese rats have consistently found that insulin levels do not change with aging [40]. Our laboratory has shown that glucose-stimulated insulin secretion from whole perfused pancreas and islets of Langerhans isolated from 6-, 12-, or 26-month-old Fischer 344 rats does not differ significantly [41, 42]. Human studies that controlled for physical activity and percent body fat demonstrated that the level of serum insulin is related to body weight rather than aging per se [43]. Although more research is needed to define the contribution of insulin to age-related body weight loss, there is currently no compelling evidence to suggest that increased insulin secretion or serum concentration underlies the mechanism leading to decreased food intake during aging.

Cholecystokinin

The gastrointestinal hormone cholecystokinin (CCK) mediates short-term satiety by acting centrally on feeding-control systems within the brain (hypothalamus) and peripherally by slowing gastric emptying and activating visceral sensory nerves. Investigation of CCK as an effecter of age-related body weight loss is based on the finding that satiation during a meal appears to occur earlier in aged vs. young humans [10, 11] and old weight-stable vs. old spontaneous-weight-loss rats [21]. While several studies have tested the hypothesis that higher levels of CCK suppress food intake to a greater degree in older vs. younger individuals, the results have not always been in agreement.

MacIntosh et al. [44] observed higher plasma CCK concentrations in older (65–80 years) versus younger (20–34 years) men during both fasting and intraduodenal lipid infusion. However, older men did report feeling satiated after nutrient infusion. These investigators concluded that there was no clear evidence that increased CCK secretion caused the reduced appetite and decreased food intake accompanying normal aging. In contrast, Khalil et al. [45] observed that comparable gallbladder contractions in response to fat ingestion in younger (22- to 42-year-old) and older (60- to 84-year-old) men and women were associated with significantly higher plasma CCK levels in the aged group, suggesting attenuated responsiveness. Consistent with the human studies, Silver et al. [46] demonstrated greater sensitivity to CCK in 25-month-old versus 8-month-old male C57BL/6 mice. That is, in response to intraperitoneal injection of an antagonist to peripheral CCK receptors, young but not old mice ate significantly more than baseline levels.

Other investigations evaluating the relationship among aging, satiety, and food intake have not supported a role for CCK. For example, basal and postprandial plasma CCK levels have been found to be similar across the ages in humans and laboratory animals [47]. In food-deprived animals, CCK reduced feeding similarly in 2-month-old and 23-month-old Wistar rats [48]. While these same investigators reported that CCK significantly reduced food intake of the oldest rats during a fixed meal, the dose of CCK needed to show differences was supraphysiological. Sandstrom and El-Salhy [49] quantified duodenal endocrine cells of NMRI mice (1, 3, 12 and 24 months old) and found no difference in the number of CCK-immunoreactive cells among the groups. Clearly, more research will be needed to better elucidate the impact that CCK may have on age-related food intake and body weight.

Conclusion

The complexity and redundancy of the physiological system controlling food intake increases the likelihood that age-related alteration to food intake and body weight can have several causes. In this regard, the available data indicate that short-term regulation of food intake via signals from the gastrointestinal tract (including CCK) is not consistently modified by aging. These short-term signals regulate individual meal size rather than body energy stores, and, thus, cannot fully explain the progressive decline in food intake and body weight observed in both humans and laboratory animals. Rather, impairments in one or more long-term signals/pathways regulating food intake are implicated. Although insulin and leptin levels are not elevated in healthy elderly humans or in old rats, alterations in the effectiveness of NPY signaling are evident in old rats experiencing spontaneous decreases in food intake and body weight, suggesting that altered NPY responsiveness is a strong candidate effecter in the alteration to energy balance late in the life span.

Much remains to be elucidated about the underlying causes of declining food intake and spontaneous weight loss near the end of life. A more thorough understanding of this phenomenon can contribute to optimizing quality of life during its final stages. Identifying the mechanisms of age-related declines in food intake and body weight may also offer clues to other aspects of the aging process.

Acknowledgements

The authors wish to thanks Jennifer Ruhe for her technical help in the preparation of this manuscript.

References

1 Stanley S, Wynne K, McGowan B, Bloom S: Hormonal regulation of food intake. Physiol Rev 2005; 85:1131–1158.
2 Morley JE, Silver AJ: Anorexia in the elderly. Neurobiol Aging 1988;9:9–16.
3 Abraham S, Carroll MD, Dresser CM: Dietary intake of persons 1–74 years of age in the United States; in Administration, HR (ed): Advance Data from Vital and Health Statistics of the National Center for Health Statistics. Rockville, Public Health Services, 1977.
4 McDowell MA, Briefel RR, Alaimo K, Bischof AM, Caughman CR, Carroll MD, Loria CM, Johnson CL: Energy and macronutrient intakes of persons ages 2 months and over in the United States: Third National Health and Nutrition Examination Survey, Phase I, 1998–1991; in Statistics, NCfH (ed): Advance Data from Vital and Health Statistics. Hyattsville, 1994.
5 Agriculture USDA, Data tables: Results from USDA's 1994–1996 diet and Health Knowledge Survey. 1997, CD-ROM, NTIS Accession Number PB98–500457.
6 Hallfrisch J, Muller D, Drinkwater D, Tobin J, Andres R: Continuing diet trends in men: the Baltimore Longitudinal Study of Aging (1961–1987). J Gerontol A Biol Sci Med Sci 1990;45:M186–M191.

7 Launer LJ, Harris T, Rumpel C, Madans J: Body mass index, weight change, and risk of mobility disability in middle-aged and older women. JAMA 1994;271:1093.

8 Roberts SB, Rosenberg I: Nutrition and aging: changes in the regulation of energy metabolism with aging. Physiol Rev 2006;86:651–667.

9 Beckoff K, MacIntosh CG, Chapman IM, Wishart JM, Morris HA: Effects of glucose supplementation on gastric emptying, blood glucose homeostasis, and appetite in the elderly. Am J Physiol Regulatory Integrative Comp Physiol 2001;280:R570–R576.

10 Rolls B, Dimeo K, Shide D: Age-related impairments in the regulation of food intake. Am J Clin Nutr 1995;62:923–931.

11 Roberts S, Fuss P, Heyman M: Control of food intake in older men. JAMA 1994;272:1601–1606.

12 Moriguti JC, Das SK, Saltzman E, Corrales A, McCrory MA, Greenberg AS, Roberts SB: Effects of a 6-week hypocaloric diet on changes in body composition, hunger, and subsequent weight regain in healthy young and older adults. J Gerontol A Biol Sci Med Sci 2000;55:B580–B587.

13 Wormsley K, Grossman M: Maximal histalog test in control subjects and patients with peptic ulcer. Gut 1965;6:427–435.

14 Bartoshuk L, Duffy V: Taste and smell; in Masoro EJ (ed): Handbook of Physiology: Aging. New York, Oxford University Press, 1995, pp 363–375.

15 Jones KL, O'Donovan D, Russo A, Meyer JH, Stevens JE, Lei Y, Keogh J, Tonkin A, Horowitz M: Effects of drink volume and glucose load on gastric emptying and postprandial blood pressure in healthy older subjects. Am J Physiol Gastro-Liver Physiol 2005;289:G240–G248.

16 Morley JE: Anorexia of aging: physiologic and pathologic. Am J Clin Nutr 1997;66:760–773.

17 Ryan AS, Elahi D: The effects of acute hyperglycemia and hyperinsulinemia on plasma leptin levels: its relationships with body fat, visceral adiposity, and age in women. J Clin Endocrinol Metab 1996;81: 4433–4438.

18 McCay CM, Crowell MF, Maynard LA: The effect of retarded growth upon the length of life span and upon the ultimate body size. J Nutr 1935;10:63–79.

19 McDonald RB, Day C, Carlson K, Stern JS, Horwitz BA: Effect of age and gender on thermoregulation. Am J Physiol Regul Integr Comp Physiol 1989;257: R700–R704.

20 McDonald RB, Florez-Duquet M, Murtagh-Mark C, Horwitz BA: Relationship between cold-induced thermoregulation and spontaneous rapid body weight loss of aging F344 rats. Am J Physiol Regul Integr Comp Physiol 1996;271:R1115–R1122.

21 Blanton CA, Horwitz BA, Geitzen DW, Griffey SM, Murtagh-Mark C, McDonald RB: Meal patterns associated with the age-related decline in food intake in the Fischer 344 rat. Am J Physiol Regul Integr Comp Physiol 1998;275:R1494–R1502.

22 McDonald RB, Hoban-Higgins TM, Ruhe RC, Fuller CA, Horwitz BA: Alterations in endogenous circadian rhythm of core temperature in senescent Fischer 344 rats. Am J Physiol Regul Integr Comp Physiol 1999;276:R824–R830.

23 Blanton CA, Horwitz BA, Blevins JE, Hamilton JS, Hernandez EJ, McDonald RB: Reduced feeding response to neuropeptide Y in senescent Fischer 344 rats. Am J Physiol Regul Integr Comp Physiol 2001;280:R1052–R1060.

24 Kalra SP, Dube MG, Sahu A, Phelps CP, Kalra P: Neuropeptide Y secretion increases in the paraventricular nucleus in association with increased appetite for food. Proc Natl Acad Sci USA 1991;88: 10931–10935.

25 Stanley BG, Anderson KC, Grayson MH, Leibowitz SF: Repeated hypothalamic stimulation with neuropeptide Y increases daily carbohydrate and fat intake and body weight gain in female rats. Physiol Behav 1989;46:173–177.

26 Kalra SP, Dube MG, Pu S, Xu B, Horvath TL, Kalra PS: Interacting appetite-regulating pathways in the hypothalamic regulation of body weight. Endocr Rev 1999;20:68–100.

27 Pich EM, Messori B, Zoli M, Ferraguti F, Marrama P, Biagini G, Fuxe K, Agnati LF: Feeding and drinking responses to neuropeptide Y injections in the paraventricular nucleus of aged rats. Brain Res 1992;575:265–271.

28 Cha CI, Lee YI, Lee EY, Park KH, Baik SH: Age-related changes of VIP, NPY and somatostatin-immunoreactive neurons in the cerebral cortex of aged rats. Brain Res 1997;753:235–244.

29 Coppola JD, Horwitz BA, Hamilton J, McDonald RB: Expression of NPY Y1 and Y5 receptors in the hypothalamic paraventricular nucleus of aged Fischer 344 rats. Am J Physiol Regul Integr Comp Physiol 2004;287:R69–R75.

30 Turenius CI, Htut MM, Prodon DA, Ebersole PL, Ngo PT, Lara RN, Wilczynski JL, Stanley BG: GABA(A) receptors in the lateral hypothalamus as mediators of satiety and body weight regulation. Brain Res 2009;1262:16–24.

31 Pu S, Jain MR, Horvath TL, Diano S, Kalra PS, Kalra SP: Interactions between neuropeptide Y and gamma-aminobutyric acid in stimulation of feeding: a morphological and pharmacological analysis. Endocrinology 1999;140:933–940.

32 Coppola JD, Horwitz BA, Hamilton J, Blevins JE, McDonald RB: Reduced feeding response to muscimol and neuropeptide Y in senescent F344 rats. Am J Physiol Regul Integr Comp Physiol 2005;288: R1492–R1498.
33 Morris DL, Rui L: Recent advances in understanding leptin signaling and leptin resistance. Am J Physiol Endocrinol Metab 2009;297:E1247–E1259.
34 Perry HM, Morley JE, Horowitz M, Kaiser FE, Miller DK, Wittert G: Body composition and age in African-American and Caucasian women: relationship to plasma leptin levels. Metabolism 1997;46: 1399–1405.
35 Wolden-Hanson T: Mechanisms of the anorexia of aging in the Brown Norway rat. Physiol Behav 2006; 88:267–276.
36 Scarpace PJ, Matheny M, Shek EW: Impaired leptin signal transduction with age-related obesity. Neuropharmacology 2000;39:1872.
37 Woods SC, D'Alessio DA: Central control of body weight and appetite. J Clin Endocrinol Metab 2008; 93:S37–S50.
38 Hill JW, Elmquist JK, Elias CF: Hypothalamic pathways linking energy balance and reproduction. Am J Physiol Endocrinol Metab 2008;294:E827–E832.
39 Wang J, Leibowitz KL: Central insulin inhibits hypothalamic galanin and neuropeptide Y gene expression and peptide release in intact rats. Brain Res 1997;777:231–236.
40 Reaven GM, Chen N, Hollenbeck C, Chen YD: Effect of age on glucose tolerance and glucose uptake in healthy individuals. J Am Geriatr Soc 1989;37:735–740.
41 Hara SL, Ruhe RC, Curry DL, McDonald RB: Dietary sucrose enhances insulin secretion of aging Fischer 344 rats. J Nutr 1992;122:2196–2203.
42 Ruhe RC, Curry DL, Herrmann S, McDonald RB: Age and gender effects on insulin secretion and glucose sensitivity of the endocrine pancreas. Am J Physiol Regul Integr Comp Physiol 1992;262: R671–R676.
43 Wang J, Ho L, Tang K, Wang L, Chen YD, Reaven G: Effect of habitual physical activity on age-related glucose intolerance. J Am Geriatr Soc 1989;37:203–207.
44 MacIntosh CG, Horowitz M, Verhagen MA, Smout AJ, Wishart J: Effect of small intestinal nutrient infusion on appetite, gastrointestinal hormone release, and gastric myoelectrical activity in young and older men. Am J Gastroenterol 2001;96:997–1007.
45 Khalil T, Walker PJ, Wiener I, Fagan CJ, Townsend CM, Greeley GH, Thompson JC: Effect of aging on gallbladder contraction and release of cholecystokinin-33 in humans. Surgery 1985;98:423–429.
46 Silver AJ, Flood JF, Song AM: Evidence for a physiological role for CCK in the regulation of food intake in mice. Am J Physiol Regul Integr Comp Physiol 1989;256:R646–R652.
47 Miyasaka K, Kanai S, Ohta M, Funakoshi A: Aging impairs release of central and peripheral cholecystokinin (CCK) in male but not in female rats. J Gerontol A Biol Sci Med Sci 1997;52A:M14–M18.
48 Voigt JP, Huston JP, Voits M, Fink H: Effects of cholecystokinin octapeptide (CCK-8) on food intake in adult and aged rats under different feeding conditions. Peptides 1996;17:1313–1315.
49 Sandstrom O, El-Salhy M: Duodenal endocrine cells in mice with particular regard to age-induced changes. Histol Histopathol 2000;15:347–353.

Roger B. McDonald, PhD
Department of Nutrition
One Shields Ave., University of California
Davis, CA 95616 (USA)
Tel. +1 530 219 5941, Fax +1 530 752 8966, E-Mail rbmcdonald@ucdavis.edu

Mobbs CV, Hof PR (eds): Body Composition and Aging.
Interdiscipl Top Gerontol. Basel, Karger, 2010, vol 37, pp 64–83

Changes in Body Composition in Response to Challenges during Aging in Rats

Tami Wolden-Hanson

Geriatric Research, Education, and Clinical Center and Research Service, Veterans Administration Puget Sound Health Care System, Seattle, Wash., USA

Abstract

Body composition changes over the lifespan of Brown Norway rats, in patterns similar to those of humans. Young adults are lean, with little fat, much of which is intra-abdominal. As they age, rats exhibit linear growth, and both lean and fat mass increase until late middle to early old age. Fat mass continues to accumulate throughout the lifespan, both viscerally and subcutaneously; aging animals carry a higher proportion of their fat mass peripherally. After middle age, skeletal muscle mass begins to decline, and sarcopenia develops when animals reach senescence. Finally, in late old age, or senescence, body weights begin to decline, and both fat and lean mass are lost. Healthy aged rats generally respond to negative energy balance challenges less robustly than younger adult animals, although they do appropriately regulate adipose tissue stores and preserve lean mass. The response to a positive energy balance challenge (high fat feeding) is less well regulated in aging animals, and dietary-induced obesity develops rapidly in aged animals. Here we present a summary of several studies of body composition in response to challenges of energy balance in aging male Brown Norway rats, with special emphasis on adipose tissue partitioning.

Aging is associated with alterations in body composition (fat and lean mass, and bone mineral content) in most mammals, including humans. These changes are often accompanied by adverse physiological consequences, as adiposity increases and lean mass decreases. Adipose tissue mass increases from middle to old age, can be associated with insulin resistance and glucose intolerance, and may progress to diabetes. Decreased skeletal muscle mass in aging can progress to sarcopenia, which results in weakness, frailty, poor immune function and impaired thermoregulation.

There are strong genetic influences upon body composition in aging; different strains of rats or mice show widely varying levels of adiposity [1–6]. Other factors that influence body composition include dietary manipulations (e.g. caloric restriction, purified diets, high-fat diets) and other environmental factors such as housing density, room temperature, and opportunity to exercise [7–11]. We have extensively studied changes in the body composition of male Brown Norway rats, a strain that does not become obese, and that is a good model for healthy human aging, and present a summary of these studies here.

Measurement of Body Composition in Aging Rodents

There are a number of methods by which body composition (fat mass and lean mass) and body fat distribution is assessed in rodents of all ages. Some methods are similar to those used in human populations (e.g. DEXA), while other methods frequently used in human populations are methodologically difficult in rodents (e.g. underwater weighing, bioelectric impedance, isotope dilution). Each method has advantages and limitations, some of which include cost of equipment, lack of precision, and expertise of operators. Each of these methods yields data on different aspects of body composition, most of which are highly correlated to each other when comparisons have been made, however none are directly comparable. Combining one or more methods yields the most information.

Dissection

Quantitative dissection of carcass fat depots, organs and muscles can be done at the time of euthanasia, or on frozen, thawed carcasses. This is a simple and cost-effective way to compare the weights of specific tissues across age or treatment groups [7].

Advantages. Dissection is easily learned, equipment costs are minimal, and results have good precision and reproducibility. This method makes it possible to directly assess sarcopenia, or age-related loss of skeletal muscle mass.

Limitations. Dissection can be performed only once on an animal, so this can be costly in terms of animal lives, if studies of changing body composition over time are needed. While it is simple to dissect and weigh organs, intra-abdominal fat depots, and specific muscles, it is not possible to dissect all fat from a carcass in order to determine whole-body adiposity, nor is it generally feasible to dissect all skeletal muscle mass.

Carcass Composition

Chemical analysis of body components has been long considered the 'gold standard' for assessment of body water, body fat and body protein [3, 12, 13]. Some methods also allow estimates of bone mass when proper equipment is available. Methodologies may differ, but carcass composition is generally performed on autoclaved carcasses, which are then homogenized. Weighed aliquots of the homogenate are dried to provide an estimate of body water, while other aliquots are subjected to solvent extraction of lipids. The main lipid extraction methods are Folch (which uses methanol and chloroform) and Soxhlet (which uses alcohols and ether), although other methods are also reported. The amount and type of fats that are retrieved from the samples may differ between methods, and chemical analysis may underestimate total carcass lipids. Some laboratories assess carcass protein content chemically, using the Kjeldahl method for measuring nitrogen, while other groups subtract calculated carcass fat from whole carcass weight for an estimate of lean mass. Bone mass may be estimated if high-heat muffle ovens are available to reduce homogenate aliquots to mineral ash.

Advantages. Carcass chemical composition can be performed in conjunction with other methods for assessment of body composition, especially dissection, because all weighed tissues can be added to the carcass before homogenization. This method has long been used in nutritional studies and is well respected as the standard for assessing levels of fat and chemical components of tissues.

Limitations. This is a labor- and time-intensive method that can be performed once only in the life of the animal. Large volumes of solvents are generated, and there are many steps for each determination. This method works best when carried out by experts, in an analytical laboratory environment that is set up to perform these types of determinations on a regular basis. New microwave technology shows promise for faster tissue extraction with lower volumes of solvents generated [14], but its use has not yet been reported in research animal carcasses.

Dual Energy X-Ray Absorptiometry (DEXA or DXA)

DEXA technology has been available since the 1990s, and provides a noninvasive method for quantification of whole body and regional body composition and bone mineral content in rats and mice [2, 15]. The methodology is based on the differential attenuation of low and high energy X-rays by the tissues in the scan area. Energy is attenuated in proportion to tissue density, and this information is used by the detector and associated software, in conjunction with tissue calibration step phantoms, to assess body composition. Soft tissues attenuate the energy beam less than bone; fat mass attenuates the beam less than lean mass. Fat mass consists primarily of adipose tissue, but lean mass includes organs, tendons, cartilage, blood, and body water in addition to skeletal muscle. The computer-driven DEXA system may also determine composition and bone mineral content in specific regions in addition to the whole body, depending upon the specific instrument and software packages used. Several instruments and manufacturers are available, including systems developed specifically for use in small (10–50 g) rodents (PIXImus, GE Lunar, Madison, Wisc., USA). Some of the instruments for use in humans (e.g. Hologics, Waltham, Mass., USA) have software packages available that allow the larger instruments to be used for rats larger than 100 g; these may not be sensitive enough for use in mice.

Advantages. This is a noninvasive measure that can be performed several times over the life of an animal, thus providing longitudinal body composition data. The scan times vary depending upon the instrument and the software, but generally take only minutes, and the data are immediately available.

Limitations. Availability of the fairly expensive instruments may be limited, or access may be limited if the instrument is normally used for human research. Animals need to be immobilized for accurate measurements, and so must be either dead or anesthetized, which can affect feeding patterns and body weights in longitudinal studies. Animal size may be an issue, especially for mice. The software makes a number of assumptions about physical properties of tissues, which may be less accurate in very small or very lean animals. One such assumption is that body water is constant, which may not be the case in aging studies.

Quantitative Magnetic Resonance Spectroscopy

A recent development for assessment of body composition in mice and rats uses magnetic resonance properties of hydrogen atoms in response to pulsed radio frequency in a stable magnetic field to determine the content of water, fat and lean mass in animals [13, 15]. Some versions of the quantitative magnetic resonance spectroscopy (QMR) also allow analysis of tissue samples up to 10 g, or biopsy samples to 30 mg. The onboard electronics in the QMR instrument (EchoMRI, Echo Medical Systems, Houston, Tex., USA) use these data to calculate whole-body water, fat and lean mass; data are immediately available. The instrument is calibrated to canola oil and gives very precise measurements of fat content. Because of the design characteristics of the instrument, lean mass is most highly correlated with skeletal muscle, and it is not clear how non-skeletal muscle lean mass such as organs, intestinal contents and skin are partitioned. The QMR instruments currently in use do not assess bone content, which is generally around 2% of total body mass. Body composition assessed by QMR is highly correlated with DEXA assessment (fig. 1), and in good agreement with carcass chemical analysis. Unanesthetized animals are placed into a cylindrical holder and inserted into a receiving port on the machine. The machine is activated using the onboard computer, and the scans are completely silent. Each scan takes less than 2 min, allowing scans in triplicate to increase precision of data collected. This instrument does not do imaging, although some groups are exploring the use of more standard magnetic resonance imaging to calculate mass of specific organs and fat depots in small subjects noninvasively [16].

Advantages. Measurements do not require the animal to be immobile, and unanesthetized animals are easily studied, although carcass measurements are also common [13]. The instrument is extremely easy to use, throughput is very fast, and measurements cause little to no stress to the animals. This method is ideal for longitudinal assessment of body composition in rats and mice of all sizes. Instruments

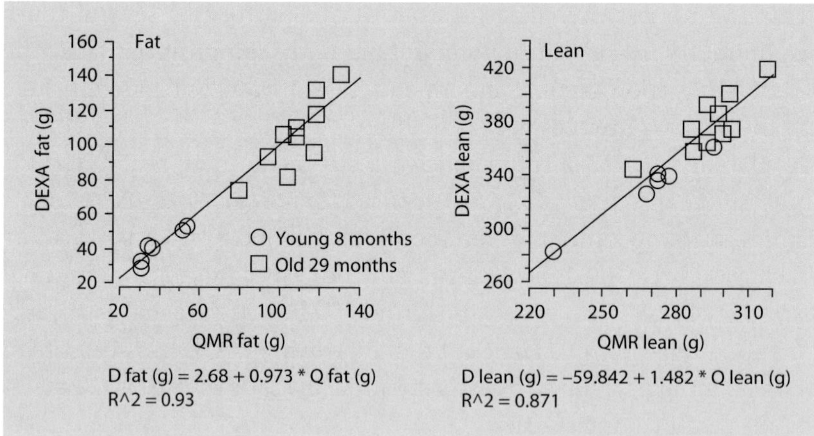

Fig. 1. Comparison of fat (left) and lean (right) mass between DEXA and QMR in carcasses of young and old Brown Norway rats, with regression equation shown below each graph. DEXA instrument was Hologics QDR4500 with Rat Whole Body software version 8.26a. QMR instrument was first generation EchoMRI (Echo Medical Systems); scans were conducted in 2004.

with the tissue and biopsy options show great promise in the postmortem assessment of sarcopenia. The use of QMR in aging studies has not yet been reported.

Limitations. Instrument availability seems to be the main limitation. While not as expensive as DEXA machines, human-sized instruments are not suitable for use in rodents, so investigators must purchase with grant money, or use in a core facility.

Body Weights in Three Rat Models of Aging

The NIA Aged Rodent Colony supplies three strains of rats (F344, Brown Norway [BN], and the FI hybrid F344BNF1) to researchers involved in aging studies. The inbred F344 is the most widely used strain in studies of aging, and the inbred BN is used the least, although the F344 strain shows more age-related pathology than either the BN or the hybrid F344BNF1 strains. Growth curves, survival curves, food intake and pathologic characterization have been published in a set of studies using large cohorts of these 3 strains of animals [1, 17]. In general, the F344 strain is shorter-lived (50% survival rate is 26mo for males) compared to the Brown Norway and the hybrid F344BNF1 (50% survival 32 and 36 months for males, respectively). Females of all strains are much smaller than the males (250–300 g); F344 females live longer than their male littermates, BN females achieve a similar age as males, and females of the hybrid strain are slightly shorter-lived. Males of the F344BNF1 strain are larger and heavier (about 600 g) than the parent, inbred strains (maximum around 500 g). Although individuals vary, population studies show that body

Fig. 2. DEXA scans (digitized and color-inverted) of representative animals from a study of cross-sectional and longitudinal changes in body composition with aging [2]. Young (5 months, left), middle-aged (14 months, center) and old (26 months, right) anesthetized male Brown Norway rats were scanned using a QDR-1500 scanner (Hologic, Inc., Waltham, Mass., USA) with Rat Whole-body Software V5.67. In this image, the color of bone is darkest and fat is lightest, with lean mass intermediate.

weights in each strain begin to decline several months before attaining the age of 50% survival.

Both the F344 and F344BNF1 strains are fatter than BN rats at comparable ages [5, 9, 10, 12, 18], as are other commonly used strains of rats, including Long-Evans [8], Wistar, and Sprague-Dawley [3, 4, 11, 16].

Patterns of Body Composition Changes with Normal Aging in Brown Norway Rats

Because of their relatively smaller body fat stores, the Brown Norway rat provides a good model for healthy human aging and has been used extensively in our laboratory [2, 19–21]. Unlike many of the fatter strains of rats, BN rats do not develop glucose intolerance or pituitary tumors, and enjoy a relatively disease-free life [1, 17]. Body composition changes over the lifespan, with patterns similar to those observed in healthy humans. We have used DEXA (fig. 2), augmented by dissection, to study longitudinal and cross-sectional changes in body composition and patterns of adiposity, in male BN rats [2]. In this study we measured body composition at two separate times: first at 5, 14 and 26 months of age, and again 3 months later at ages 8, 17 and 29 months, with a fourth group of very young animals (3 months of age) added at that time. As they age, rats exhibit linear growth, as can be seen in figure 2, and both

in the long-term study of body composition (fig. 3), these data indicate that refeeding after food deprivation preferentially shifted accumulation of fat to the intra-abdominal depots (fig. 6c), although again the effect was strongly related to age, and greatest in the young animals.

Lean Mass

DEXA quantifies a number of non-skeletal muscle components as lean mass, including blood, intestinal contents and organs. Intestinal contents of fasted rats would be essentially nil, whereas a large rat can contain 20 g or more of food in varying stages of absorption. Thus, it is not necessarily skeletal muscle that is being lost during such studies of fasting. We dissected quadriceps and gastrocnemius muscles from both hind limbs and show (fig. 7a, b) that food deprivation did not decrease muscle weights, although there was a strong effect of age on muscle weights, and senescent rats do show signs of significant sarcopenia.

Organ weights (liver, heart and kidneys) were significantly increased with age, with the largest change between 16 and 26 months. Not surprisingly, fasting significantly decreased (~20%) the weights of these organs (fig. 7c, d), primarily due to effects upon the liver.

Changes in Body Composition in Response to Positive Energy Balance

The obesity epidemic currently plaguing Westernized societies is likely due to easy availability of high-fat diets, combined with a lack of physical activity. We fed young and old male Brown Norway rats regular chow (Purina 5001; 3.2 kcal/g) or a high fat diet (Research Diets D12266B; 32% kcal from fat; 4.4 kcal/g) for 8 weeks, then subjected them to 72 h food deprivation, followed by 4 weeks of re-feeding (fig. 8) [Wolden-Hanson and Matsumoto, unpubl. data].

Body Weights

Both young and old high fat-fed rats were hyperphagic initially, and gained weight rapidly, but young rats soon reduced their volume of food intake, while old animals continued their hyperphagia, nearly doubling their caloric intake (data not shown). Weight loss resulting from 72 h of food deprivation was blunted in old high fat-fed rats, compared to chow-fed old rats (fig. 8), but was not different in young rats. Re-feeding restored body weights of high fat-fed young and old rats to >100% of prefasted levels by the end of the 28-day re-feeding period, while the chow-fed old rats did not return to prefasted levels, but did match body weights with nonfasted controls.

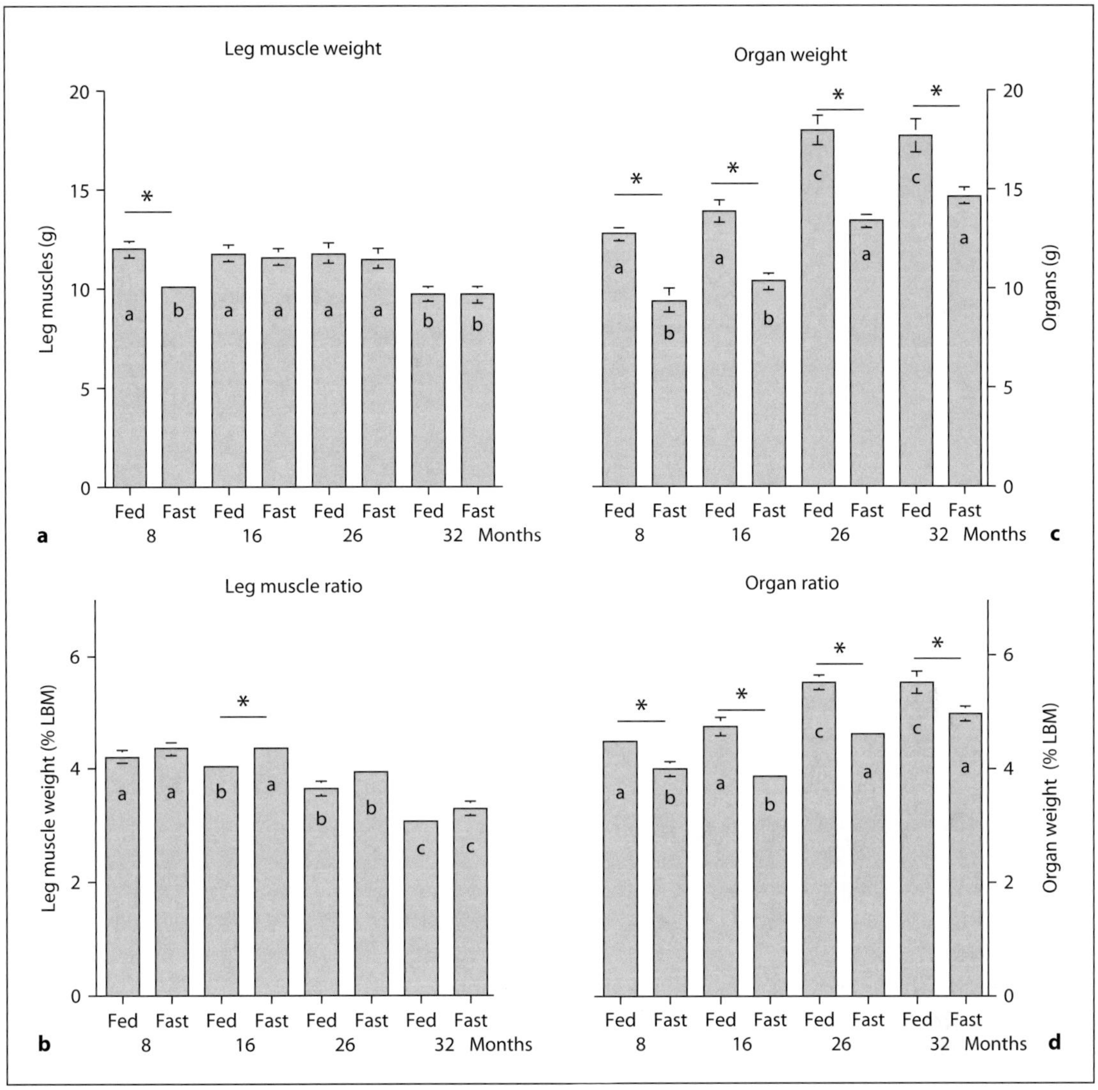

Fig. 7. Skeletal muscle and organ mass in young (8 months), middle-aged (16 months), old (26 months) and senescent (31–32 months) male BN rats comparing the effects of 72 h food deprivation (fast) and 12 days re-feeding (fed). **a** Leg muscle weight (g) is the sum of weights of dissected gastrocnemius and quadriceps muscles from both hind limbs. **b** Leg muscle ratio is calculated as leg muscle weight divided by lean mass ×100 for %. **c** Organ weight (g) is the sum of weights of dissected liver, heart and kidneys. **d** Organ ratio is calculated as organ weight/lean mass × 100 for %. For this set of figures, data in columns with the same letter are not significantly different from each other. Bars with * are significantly different ($p < 0.05$) between fasting conditions within age group. n = 7–12 per age and treatment group.

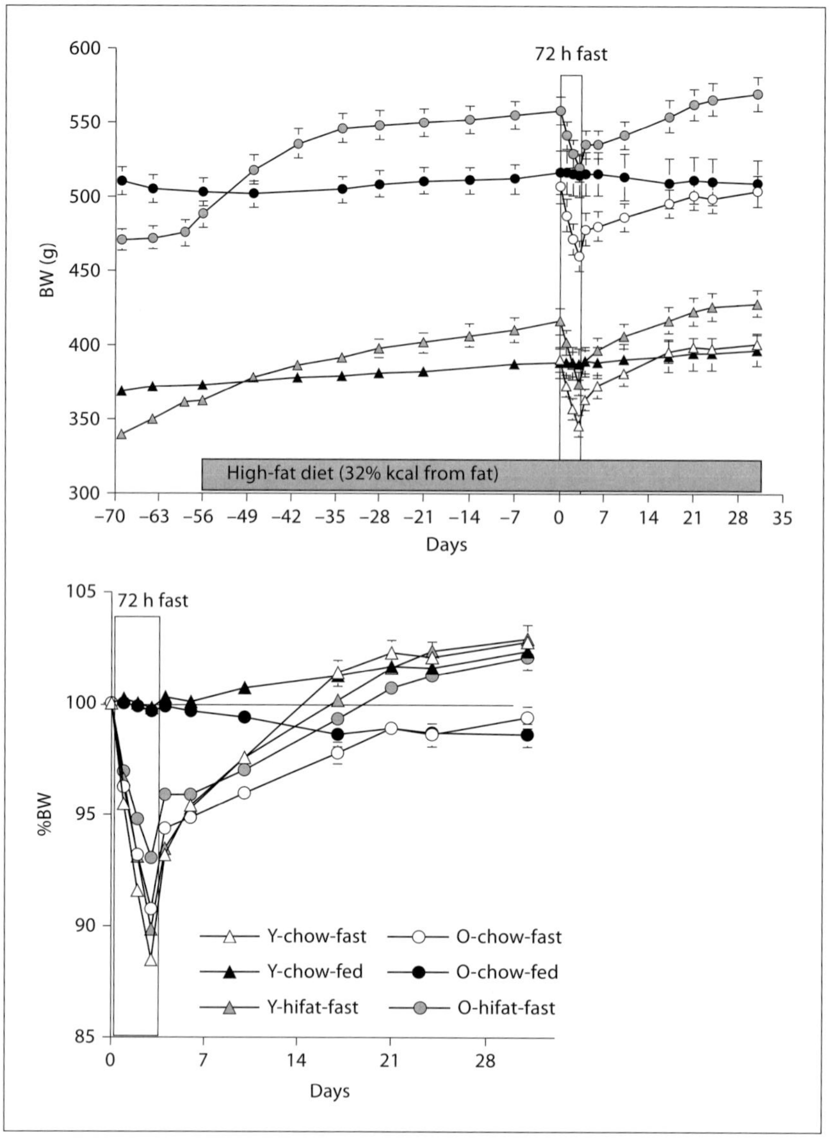

Fig. 8. Body weights of young and old male Brown Norway rats fed chow (Purina 5001) or high-fat diet (32% kcal from fat; Research Diets D12266B) for 8 weeks, subjected to 72 h food deprivation, and then re-fed for 4 weeks (final ages 8 and 29 months; n = 7–10/group). Weights are expressed as absolute (g; upper) and as percentage of pre-fast 3 days average (%BW; lower). Body weights of fasted chow-fed rats of both ages recovered to the body weights of non-fasted chow-fed cohorts by the end of the 28 days re-feeding period [Wolden-Hanson and Matsumoto, unpubl. data].

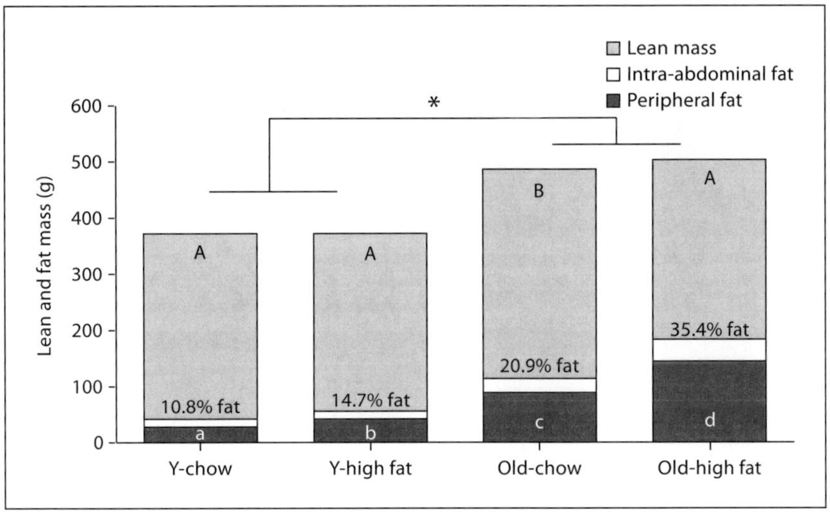

Fig. 9. Body composition of HiFat (32% kcal from fat) and chow-fed young (8 months) and old (29 months) male Brown Norway rats. Total fat and lean mass (g) were assessed on carcasses by DEXA (Hologics QDR4500). Intra-abdominal fat (white bars) was dissected and weighed, then returned to carcass for DEXA analysis. Peripheral fat is the difference between total and intra-abdominal fat. Data in columns with the same letter are not significantly different from each other; capital letters relate to lean mass, lower case letters relate to total fat mass. Intra-abdominal fat mass was different between all groups except young. Body weights were significantly different with age, but not with diet (* $p < 0.05$) [Wolden-Hanson and Matsumoto, unpubl. data].

Body Composition

Body composition analysis of fat and lean mass (fig. 9) was accomplished by dissection of intra-abdominal fat, followed by DEXA (Hologics QDR4500). Dissected fat depots were weighed and then returned to the carcass for DEXA evaluation of total lean and fat mass. Dissected intra-abdominal fat weight (the sum of all intra-abdominal fat depot weights) was subtracted from total fat to calculate peripheral fat mass. Because fasted chow-fed animals had returned to the weights of their nonfasted controls, and their results were not different from non-fasted animals, these groups were combined in further analyses. Body weights of young chow-fed animals were not different from high fat-fed animals 28 days after food deprivation, nor were lean mass and intra-abdominal fat mass. Peripheral fat mass was slightly but significantly increased in high fat-fed young animals, and this could be detected during normal handling of the animals. In the case of old animals, although body weights were not significantly different from chow-fed rats, each of the 3 compartments (lean mass, intra-abdominal fat and peripheral fat mass) was significantly different in the high fat group. Old rats on high fat diet had decreased lean mass levels that were comparable to that of the younger, smaller animals. Intra-

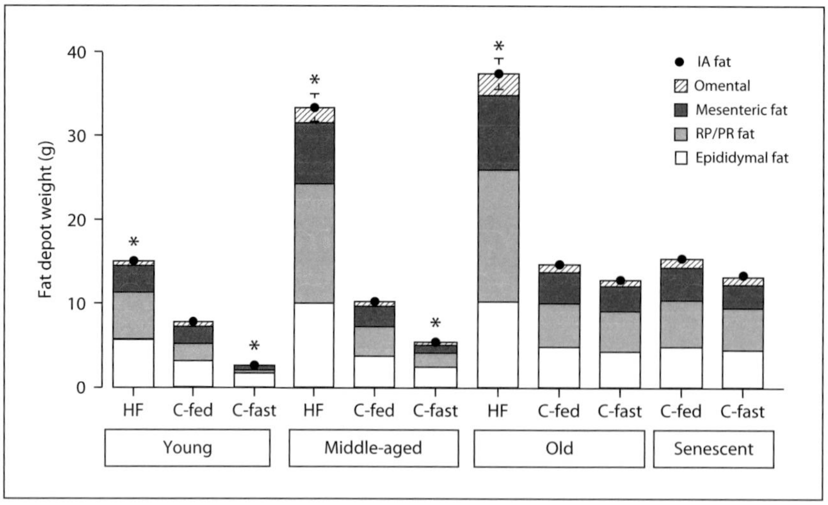

Fig. 10. Weights of individual intra-abdominal fat pads in animals on high-fat diet (HF) compared with chow-fed (C-fed) and 72 h food deprived (C-fast) young (8 months), middle-aged (17 months), old (26 months) and senescent (32 months) rats. Intra-abdominal fat (IA; sum of all fat pads), includes omental fat, mesenteric fat, retroperitoneal/peri-renal fat (RP/PR fat) and epididymal fat. * $p < 0.05$ compared to C-fed rats within the age groups.

abdominal fat was doubled in old high fat-fed rats, and peripheral fat mass was nearly doubled (fig. 9).

Fat Depot Weights

There are 4–5 commonly recognized intra-abdominal fat pads. These include: the paired epididymal (Epi) pads, which surround the testes and extend rostrally; the retroperitoneal and perirenal depots (here combined as RP/PR), which are paired, and located along the dorsal surface of the peritoneal cavity, the rostral portion of which surrounds the kidneys; the single mesenteric (Mes) depot, which is attached to the gastrointestinal tract; and the small omental depot, which is located between the spleen and stomach, and is closely associated with the pancreas. Some investigators combine the mesenteric and omental fat pads into a single measure of visceral adiposity, whereas others may refer to all intra-abdominal fat as visceral fat.

To gain insight into the effect of positive and negative energy balance upon these individual intra-abdominal fat depots, we compared (fig. 10) the weights of each fat pad in animals fed the high fat diet (described below) with those of chow-fed (C-fed) and 72 h food deprived (C-fast) rats described in section 5.3. There were no senescent rats in the high fat diet study. Intra-abdominal fat weights (IA; sum of all fat pads) were significantly increased by high fat diet at each age studied, compared to chow-fed rats. High fat diet doubled the total IA fat mass in young and old rats, and increased it 3.3-fold in middle-aged rats (IA fat in middle-aged rats was not different from old). This increase in IA fat was reflected in increases of all fat depots at all ages studied, with the exception of the omental fat pad in young animals. The depot with greatest fold increase in young rats was the retroperitoneal/peri-renal pad (RP/PR; 2.7-fold), while both the omental and RP/PR depots increased by 4-fold in middle-aged and by 3-fold in old rats.

Food deprivation for 72 h significantly decreased IA fat mass in young (to 35% of fed) and middle-aged (to 54% of fed), but not old or senescent, chow-fed rats (fig. 10). The weights of all fat depots, except the omental fat pad, were decreased significantly in both young and middle-aged rats. The weights of the mesenteric and retroperitoneal/peri-renal fat pads were decreased to 21–24% of fed in young, and to 43–44% in middle-aged animals. The epididymal fat pad was less affected; it weighed 53 and 68% of fed young and middle-aged animals, respectively. In old and senescent rats, the only fat pad weight that was significantly affected was the mesenteric, which was decreased to 76–79% of the weight of the same fat depot in fed rats.

These differential responses to energy balance challenges by the individual fat depots in different age groups show the importance of dissection in conjunction with measurement of whole-animal body composition. Many investigators weigh only the easily accessible epididymal fat pad as their only measure of the effect of treatments upon adiposity, and it is clear from the data presented here that this is not the best route to take.

Conclusions

Taken together, these studies have shown that, in studies of aging, comparison of body weights alone will not give an accurate assessment of energy flux due to treatments or challenges to energy balance. Although body weights may plateau in late middle to early old age, old and senescent rats are fatter than younger animals; the fat they bear intra-abdominally is resistant to lipolysis resulting from situations of negative energy balance such as food deprivation, but is capable of increasing at rates similar to young rats in situations of positive energy balance. The large pool of peripheral adipose tissue may contribute more to maintenance of energy balance than commonly recognized. Whole-body lean mass, as measured by DEXA, is not a marker for sarcopenia, because it includes non-skeletal muscle tissues that increase with the size, weight and age of the animal; the only good measure for sarcopenia is direct dissection of skeletal muscles, not all of which are affected by age. However, because the lean mass assessment by QMR is said to be more directly reflective of skeletal muscle than other whole-body methodologies [13, 15], it may be that QMR will accurately measure sarcopenia, however this remains to be tested. To properly assess body composition in rats of different ages, it is advisable to combine whole body measures (by

QMR, DEXA or carcass composition analysis) with assessment of specific components, either by dissection or possibly by MRI [16].

Acknowledgement

This work was supported exclusively by a VA Merit award to Dr. Wolden-Hanson and by use of resources and facilities at VA Puget Sound Health Care System, Seattle Division.

References

1 Turturro A, Witt WW, Lewis S, Hass BS, Lipman RD, Hart RW: Growth curves and survival characteristics of the animals used in the biomarkers of aging program. J Gerontol A Biol Sci Med Sci 1999; 54:B492–B501.

2 Wolden-Hanson T, Marck BT, Smith L, Matsumoto AM: Cross-sectional and longitudinal analysis of age-associated changes in body composition of male Brown Norway rats: association of serum leptin levels with peripheral adiposity. J Gerontol A Biol Sci Med Sci 1999;54:B99–B107.

3 Klinger MM, MacCarter GD, Boozer CN: Body weight and composition in the Sprague Dawley rat: comparison of three outbred sources. Lab Anim Sci 1996;46:67–70.

4 Lesser GT, Deutsch S, Markofsky J: Aging in the rat: longitudinal and cross-sectional studies of body composition. Am J Physiol 1973;225:1472–1478.

5 Rikans LE, Venkataraman PS, Cai Y: Evaluation by dual-energy x-ray absorptiometry of age-related changes in body composition of male rats. Mech Ageing Dev 1993;71:155–158.

6 West DB, Boozer CN, Moody DL, Atkinson RL: Dietary obesity in nine inbred mouse strains. Am J Physiol 1992;262:R1025–R1032.

7 Nolen GA: Effect of various restricted dietary regimens on the growth, health and longevity of albino rats. J Nutr 1972;102:1477–1493.

8 Garthwaite SM, Cheng H, Bryan JE, Craig BW, Holloszy JO: Ageing, exercise and food restriction: effects on body composition. Mech Ageing Dev 1986;36:187–196.

9 McDonald RB, Carlson K, Day C, Stern JS, Horwitz BA: Effect of gender on the response to a high fat diet in aging Fischer 344 rats. J Nutr 1989;119:1472–1477.

10 Olfert IM, Balouch J, Mathieu-Costello O: Oxygen consumption during maximal exercise in Fischer 344 x Brown Norway f1 hybrid rats. J Gerontol A Biol Sci Med Sci 2004;59:801–808.

11 Skalicky M, Narath E, Viidik A: Housing conditions influence the survival and body composition of ageing rats. Exp Gerontol 2001;36:159–170.

12 Feely RS, Larkin LM, Halter JB, Dengel DR: Chemical versus dual energy X-ray absorptiometry for detecting age-associated body compositional changes in male rats. Exp Gerontol 2000;35:417–427.

13 Nixon JP, Zhang M, Wang C, Kuskowski MA, Novak CM, Levine JA, Billington CJ, Kotz CM: Evaluation of a quantitative magnetic resonance imaging system for whole body composition analysis in rodents. Obesity (Silver Spring) 2010;in press.

14 Pare JR, Matni G, Belanger JM, Li K, Rule C, Thibert B, Yaylayan V, Liu Z, Mathe D, Jacquault P: Use of the microwave-assisted process in extraction of fat from meat, dairy, and egg products under atmospheric pressure conditions. J AOAC Int 1997;80: 928–933.

15 Tinsley FC, Taicher GZ, Heiman ML: Evaluation of a quantitative magnetic resonance method for mouse whole body composition analysis. Obes Res 2004;12:150–160.

16 Tang H, Vasselli JR, Wu EX, Boozer CN, Gallagher D: High-resolution magnetic resonance imaging tracks changes in organ and tissue mass in obese and aging rats. Am J Physiol Regul Integr Comp Physiol 2002;282:R890–R899.

17 Lipman RD, Chrisp CE, Hazzard DG, Bronson RT: Pathologic characterization of Brown Norway, Brown Norway x Fischer 344, and Fischer 344 x Brown Norway rats with relation to age. J Gerontol A Biol Sci Med Sci 1996;51:B54–B59.

18 McCarter RJ, Palmer J: Energy metabolism and aging: a lifelong study of Fischer 344 rats. Am J Physiol 1992;263:E448–E452.

19 Matsumoto AM, Marck BT, Gruenewald DA, Wolden-Hanson T, Naai MA: Aging and the neuroendocrine regulation of reproduction and body weight. Exp Gerontol 2000;35:1251–1265.

20 Wolden-Hanson T, Marck BT, Matsumoto AM: Troglitazone treatment of aging Brown Norway rats improves food intake and weight gain after fasting without increasing hypothalamic *npy* gene expression. Exp Gerontol 2002;37:679–691.

21 Yukawa M, Weigle DS, Davis CD, Marck BT, Wolden-Hanson T: Peripheral ghrelin treatment stabilizes body weights of senescent male Brown Norway rats at baseline and after surgery. Am J Physiol Regul Integr Comp Physiol 2008;294: R1453–R1460.

22 Wolden-Hanson T, Marck BT, Matsumoto AM: Blunted hypothalamic neuropeptide gene expression in response to fasting, but preservation of feeding responses to AgRP in aging male Brown Norway rats. Am J Physiol Regul Integr Comp Physiol 2004; 287:R138–R146.

23 Sohn EH, Wolden-Hanson T, Matsumoto AM: Testosterone (T)-induced changes in arcuate nucleus cocaine-amphetamine-regulated transcript and npy mrna are attenuated in old compared to young male Brown Norway rats: Contribution of T to age-related changes in cocaine-amphetamine-regulated transcript and *npy* gene expression. Endocrinology 2002;143:954–963.

24 Gruenewald DA, Marck BT, Matsumoto AM: Fasting-induced increases in food intake and neuropeptide y gene expression are attenuated in aging male Brown Norway rats. Endocrinology 1996;137: 4460–4467.

25 Wolden-Hanson T: Mechanisms of the anorexia of aging in the Brown Norway rat. Physiol Behav 2006; 88:267–276.

26 McDonald RB, Florez-Duquet M, Murtagh-Mark C, Horwitz BA: Relationship between cold-induced thermoregulation and spontaneous rapid body weight loss of aging F344 rats. Am J Physiol 1996; 271:R1115–R1122.

Tami Wolden-Hanson, PhD
VA Puget Sound Health Care System (S-182-GRECC)
1660 South Columbian Way
Seattle WA 98108–1597 (USA)
Tel. +1 206 768 5394, Fax +1 206 764 2569, E-Mail twh@u.washington.edu

Mobbs CV, Hof PR (eds): Body Composition and Aging.
Interdiscipl Top Gerontol. Basel, Karger, 2010, vol 37, pp 84–93

New Haystacks Reveal New Needles: Using *Caenorhabditis elegans* to Identify Novel Targets for Ameliorating Body Composition Changes during Human Aging

Catherine A. Wolkow

Laboratory of Neurosciences, NIA Intramural Research Program, NIH, Baltimore, Md., USA

Abstract

Dramatic changes in body composition accompany aging in humans, particularly with respect to adiposity and the musculature. People accumulate fat as they age and lose muscle mass and strength. *Caenorhabditis elegans* nematodes are small, hermaphroditic soil nematodes that offer a flexible model for studying genetic pathways regulating body composition in humans. While there are significant physiological differences between worms and people, many of the genetic pathways relevant to human lipid and muscle homeostasis are present in worms. Initial studies indicate that adiposity increases in *C. elegans* during aging, as occurs in humans. Furthermore, substantial evidence demonstrates age-related loss of muscle mass in worms. Possible mechanisms for these changes in *C. elegans* are presented. Recent studies have highlighted neuroendocrine and environmental signals regulating *C. elegans* fat metabolism. Potential dysfunction of these pathways during aging could affect overall fat accumulation. By contrast, muscle decline in aging worms results from accumulated damage and 'wear-and-tear' over life span. However, neuroendocrine pathways also regulate muscle mass in response to food availability. Such pathways might provide useful therapeutic approaches for combating muscle loss during aging. From this chapter, readers will develop a deeper understanding of the ways that *C. elegans* can be used for mechanistic gerontological studies.

Introduction to *Caenorhabditis elegans*

C. elegans *as a Model for Gerontological Studies*

C. elegans nematodes are small soil nematodes that have been adopted by researchers as a simple experimental model for investigating physiological and developmental processes in a whole organism [1]. There are several characteristics that make *C. elegans* a powerful experimental approach for gerontology. One, the body is transparent and the tissues can be easily visualized through a microscope in live animals. This allows visual inspection for aging-related anatomical changes. Two, worms are small, have a short life cycle and can be maintained economically on agar-based medium in Petri dishes, without any specialized equipment. These features facilitate age-synchronization of large populations for longitudinal or cross-sectional studies, providing statistical robustness. Three, *C. elegans* are amenable to a variety of experimental manipulations facilitating mechanistic studies. Investigations commonly involve genetic analysis, through forward and reverse genetic approaches, which are complemented by cellular and behavioral studies.

Genetic studies have been tremendously successful for identifying new pathways regulating *C. elegans* longevity [2]. In many cases, genes controlling longevity also impact tissue aging, such that mutants maintain high functionality late into life [3]. Genes with this phenotype activate stress-resistance programs, for example the insulin/IGF-I pathway which regulates expression of heat-shock proteins and antioxidant enzymes [4–6]. Similarly, the antioxidants regulated by the SKN-1/Nrf family transcription factor also promote longevity and stress resistance [7]. This link between stress resistance and longevity suggests that tissue decline during *C. elegans* aging results at least in part from tissue damage, potentially by oxidative insult and proteotoxicity.

Regulation of Adiposity in C. elegans

Evidence for Fat Accumulation during Aging in C. elegans

Worms lack specialized lipid storage cells, but accumulate fat in the intestine and hypodermis [9]. The *C. elegans* intestine serves digestive, detoxification and metabolic storage functions, thus combining the functions of the human intestine, liver and adipocyte [8]. Under standard growth conditions, wild-type adults contain moderate amounts of fat in the intestine and hypodermis that can be visualized by Sudan black or oil red O staining [9, 10]. These depots constitute the major metabolic fat stores and are depleted during starvation [10].

Only one study has closely examined adiposity during aging in *C. elegans* [11]. Using two different fat-specific stains, Nile Red and Sudan Black, fat levels were found to increase over life span in wild-type animals. Most of this increase occurred between adult days 5 and 10, which is the period following reproductive senescence when animals also exhibit slowing locomotory behaviors [12, 13]. Thus, increased fat accumulation corresponds with sedentary behavior and possibly reflects a metabolic shift related to cessation of egg laying and locomotion. However, mechanisms contributing to age-related adiposity have not been elucidated. Therefore, the following sections discuss the current understanding of *C. elegans* fat accumulation and metabolic regulation. Shifts in these pathways during aging could be factors contributing to increased adiposity in old worms.

Function and Regulation of Fat Storage in C. elegans

Fat serves as a source for metabolic energy for worms under several conditions. First, there is a requirement for basal fat metabolism during growth and development. Mutant worms unable to synthesize certain fats fail to develop normally or die prematurely, suggesting that fat performs essential functions for development and survival [14, 15]. In adults, intestinal fat is mobilized to oocytes where it provides nutrients to developing embryos [16, 17]. Finally, in adults and diapaused dauer larvae, stored fat serves as an energy source in the absence of environmental food sources [18]. Using fat-specific dyes, fat depletion can be observed as animals endure starvation conditions [14].

Despite differences in the sites of fat storage between mammals and invertebrate nematodes, several of the genetic factors that regulate mammalian adiposity have also turned up as significant regulators of *C. elegans* fat metabolism [19]. In mammals, fat storage is regulated by transcription factors that specify adipocyte cell fate and then induce the expression of adipocyte-specific fat storage enzymes. Mammalian adipocyte differentiation and adipogenesis are regulated by the C/EBP transcription factor. A C/EBP homolog in worms, LPD-2, regulates the expression of lipogenic genes [14]. A second transcription factor, SREBP, is critical in mammals for lipid homeostasis. Similarly, a *C. elegans* SREBP homolog, LPD-1/SBP-1, also regulates lipogenic gene expression [14]. Expression of *fat-5, fat-6* and *fat-7*, which encode fatty acid desaturases, is also under *lpd-1* control. Mutants lacking *lpd-1*/SREBP function show an absence of intestinal fat and growth defects, suggesting that fatty acids are necessary for normal development [14].

Nuclear hormone receptors regulate fatty acid metabolism in both worms and people. A central regulator of mammalian lipid metabolism is HNF4, a member of the steroid nuclear receptor transcription factor family. Similarly, fatty acid metabolism in *C. elegans* is regulated by two HNF4-like receptors, encoded by the *nhr-49* and *nhr-80* genes [20, 21, 15]. Both NHR-49 and NHR-80 regulate the expression of delta-9 desaturases (stearoyl-CoA desaturases). The *C. elegans* genome contains three delta-9 desaturase genes, *fat-5*, *fat-6* and *fat-7*. Deletion of all three delta-9 desaturases is lethal, demonstrating the necessity of the lipid products for survival [15]. While these three desaturase enzymes function redundantly, *fat-6* appears to encode the major delta-9 desaturase active under standard growth conditions [15]. Under certain conditions, such as when *fat-6* activity is limited, *fat-7* expression is highly induced and this induction involves *nhr-80* [15]. In contrast, *nhr-49* activity is necessary for *fat-7* expression to be repressed under starvation conditions [21]. While both NHR-49 and NHR-80 regulate delta-9 desaturase expression, NHR-49 has additional roles regulating the expression of genes involved in beta-oxidation and other starvation-regulated genes [20, 21]. An important cofactor for both NHR-49 and LPD-1/SBP-1 is the mediator protein, MDT-15 [22]. In the absence of *mdt-15* activity, *nhr-49*-mediated induction of FA desaturases during starvation is compromised. Consistent with this, levels of unsaturated fatty acids are elevated in *mdt-15* mutants. Mutations in either *nhr-49* or *mdt-15* significantly shorted adult life span, demonstrating that fatty acid metabolism is an essential aspect of *C. elegans* physiology [20, 22]. Further evidence that fats play important roles in *C. elegans* physiology is the finding that elongase mutants develop into sickly adults with low fertility, or arrest development completely in the first larval stage [23].

Neuroendocrine and Environmental Factors Promote Mobilization of Fat Reserves
Fat metabolism requires the action of lipases, esterases, enzymes of beta-oxidation and the mitochondria electron transport chain. When environmental nutrients are depleted, fat reserves in the intestine and hypodermis provide metabolic substrates for growth and survival. Neuroendocrine pathways regulate fat utilization under these conditions. In adults, sensory neurons detecting food secrete a TGF-β ligand, DAF-7, into the body [24]. Low levels of food in the environment correspond to reduced *daf-7* expression. Low *daf-7* expression stimulates fat accumulation in the intestine and hypodermis [25]. Environmental food signals also increase serotonin levels, which, in turn, inhibits fat accumulation and serotonin-deficient mutants accumulate high levels of fat [26]. Serotonin signaling also regulates the rate of fat metabolism in peripheral tissues [27]. It is unlikely that a simple decline in serotonin signaling is the cause for fat accumulation in older worms. In addition to regulating fat accumulation, serotonin also regulates feeding and locomotory behaviors, several of which are altered during aging [28, 29]. Mutations disrupting serotonin receptors can suppress these age-related changes, suggesting that serotonergic neurotransmission is elevated, rather than depressed, during normal aging. Thus, the mechanisms driving fat accumulation during worm aging may occur downstream of serotonergic input.

Increased utilization of fat reserves can have a prolongevity effect in *C. elegans*. Both fat mobilization and longevity are enhanced by disruption of *ctbp-1*, which encodes a homolog of the mammalian NAD(H)-dependent transcriptional co-repressor, C-terminal binding protein [30]. In *ctbp-1* mutants, triacylglycerol levels are reduced due to the induction of *lips-7*, encoding a TAG lipase. Interestingly, *ctbp-1* mutants are long-lived due, in part, to the induction of *lips-7*.

Defects in mitochondrial function caused by prohibitin mutations differentially affect longevity of wild-type and long-lived mutants. Prohibitins are conserved components of mitochondrial membranes that are required for mitochondrial function, which is also involved in fatty acid metabolism. *C. elegans* prohibitin mutants are unable to metabolise fat, and have substantial decreases in stored fats [11]. This fat storage defect was enhanced in long-lived *daf-2*/insulin-receptor mutants. Although prohibitin mutations shortened life span in a wild-type animals, they were found to further extend life span of long-lived insulin receptor mutants. These findings suggest a complex interaction between prohibitin-dependent aspects of mitochondrial function and fat metabolism in *C. elegans*.

AMPK, a multisubunit kinase regulated by cellular energy levels, coordinates the breakdown of intestinal and hypodermal fat with energy requirements of the periphery. Dauer larvae lacking adequate AMPK activity in hypodermal cells prematurely exhaust their lipid stores and die [18]. The excretory cell constitutes the major metabolic sink required for dauer survival, and AMPK mutant dauers die from osmotic stress due to inadequate excretory cell function. This phenotype reflects inappropriately high activity of the AMPK target, ATLG-1, an adipose triglyceride lipase, in the AMPK mutant dauers. Since AMPK disruption increases fat utilization and depletes fat reserves, it is possible that fat accumulation during aging reflects elevated AMPK activity.

Intestine as a Central Regulator of C. elegans *Longevity*

Changes in intestinal composition during aging may reflect alterations in the function of aging-regulated transcription factors. This idea is supported by findings from microarray analyses of transcription during aging and in long-lived mutants which found that a substantial number of age-regulated transcripts are targets of the ELT-3/GATA transcription factor, which is expressed in the worm intestine and hypodermis [31]. Indeed, the group of age-regulated transcripts identified by microarray analysis was enriched for intestinal genes. It is possible that changes in body composition in humans might also reflect tissue-specific alterations in gene expression due to age-regulated transcription factors.

The intestine plays a central role in determining *C. elegans* life span. This is due to the fact that the FOXO transcription factor, DAF-16, appears to function primarily from the intestine to promote longevity [32]. DAF-16 directs the expression of prolongevity genes with a variety of functions, such as antioxidant defense, xenobiotic detoxification and fatty acid metabolism [33, 34]. DAF-16 was first identified as the major target of DAF-2, an insulin/IGF-I receptor like protein that regulates *C. elegans* development, reproduction and longevity [35, 36]. One possibility is that DAF-16 target genes protect the intestine from aging-related decline, allowing the intestine to carry on essential functions for longer time. In addition, there may be DAF-16-regulated changes in intestinal function that have prolongevity effects. One example is a shift to fat-based metabolism. The *C. elegans* insulin pathway regulates the expression of a large number of transcripts, either directly or indirectly through *daf-16* [34]. Among these insulin-pathway target genes is a lipase, corresponding to the *K04A8.5* open reading frame [37]. Inactivation of *K04A8.5* function partially suppressed the longevity of insulin receptor mutant adults, suggesting that utilization of fat stores via K04A8.5 lipase activity was partially responsible for the extreme longevity of adults with defective insulin signaling.

Muscle Aging and Sarcopenia in *C. elegans*

Sarcopenia in Feeding and Locomotory Muscles

In addition to increased adiposity, loss of muscle mass is a major body composition change of human aging. This loss of muscle mass is named sarcopenia and the causes remain mysterious, although they are likely to be complex in the human population. Clinical and laboratory studies of sarcopenia have implicated a variety of factors, including disruptions in neuromuscular connectivity, immune dysfunction, and mitochondrial dysfunction [38, 39]. Changes in musculature similar to human sarcopenia have been noted in *C. elegans* locomotory and feeding muscles [13, 40–42]. Identifying the factors driving muscle sarcopenia during worm aging would identify candidate pathways involved in human sarcopenia and potentially lead to the development of new therapeutic approaches for its treatment.

The locomotory muscles are situated longitudinally along the body axis in four groups and function as pairs to promote sinusoidal movement along the substrate. Sarcomeres in the *C. elegans* locomotory muscles degenerate as animals age, and this damage can be visualized by GFP-labeled sarcomeric proteins, phalloidin-stained sarcomeres or ultrastructurally by electron microscopy [13, 41]. Prior to the appearance of sarcomeric damage, however, animals exhibit locomotory deficiencies that reflect weakness in the muscle sarcomeres [12, 13, 41]. At this stage of aging, when the muscles appear generally intact but locomotion has slowed, muscles become more susceptible to damage [13]. In long-lived insulin pathway mutants, both muscle damage and locomotory declines are delayed [12, 41]. This correlates with the induction of stress resistance enzymes that could protect muscles from damage over life span [5, 34, 43].

Sarcopenia has also been studied in the *C. elegans* pharynx, which is the neuromuscular feeding organ in the head. In contrast to the longitudinal muscles promoting locomotion, the pharynx muscles are radially organized and contract to crush the food, which consists of bacteria and slime molds [44]. Damage to the pharynx muscles is more readily apparent than in the body muscles and has been visualized by Nomarski optics in live animals, but can also be seen by GFP-marked muscle components [3]. There is evidence to suggest that pharynx deterioration results directly from damage accumulated over a lifetime of contractions, as pharynx function was protected in animals that were prevented from fast contractions for several days [42].

Functional decline of the feeding (pharynx) and locomotion (body) muscles is correlated with life span [12]. Individual animals that maintained longer spans of high pharynx or body muscle function were more likely to have longer life span, and vice versa. In these studies, there was no correlation between self-fertile reproductive period and life span showing that progeny production in these hermaphrodites was not a measureable predictor of relative longevity. A computer learning-based analysis of pharynx muscle morphology across adult life span revealed that morphology changes during aging occurred in a stepwise, rather than progressive, fashion [45]. This revealed that structural aging may be modulated by cellular homeostatic mechanisms for maintaining cellular integrity.

Starvation Alters Muscle Composition in Ways that Mimic Aging

As for human muscles, the causes of sarcopenia in worm muscles remain mysterious. One possibility is a connection to the decline in feeding that occurs during normal aging [12]. Evidence of feeding decline is a progressive decrease in the rate of food intake by the pharynx muscles as well as atrophic changes in the intestine. Thus, a potential approach to studying sarcopenia in worms is through the starvation response in young adults. Such studies have shown that one response to starvation in pharynx muscles is an increase in muscarinic signaling [46]. Increased pharynx muscarinic signaling leads to activation of the MPK-1 kinase, which is a worm ortholog of the mammalian stress-inducible kinase ERK. This is believed to prime the pharynx muscles to quickly activate pumping upon future encounters with food. Overactive muscarinic activation during starvation can have detrimental consequences. For example, *gbp-2* mutants that are unable to contain muscarinic signaling due to the absence of an inhibitory G-protein subunit, have significantly shortened survival times under starvation conditions [46, 47].

One output of starvation-induced muscarinic signaling in the pharynx is activation of the autophagy pathway. Autophagy is a process that enables cells to recycle components through lysosomal proteolysis to obtain nutrients [48]. There is evidence suggesting that autophagy is also activated in mouse models of sarcopenia [49]. The starvation-sensitive *gbp-2* mutants demonstrate overactive autophagy in the pharynx [50]. Autophagic overactivity is one cause of the early death in *gbp-2* mutants, as genetic manipulations that depress autophagy can partially rescue the survival of starved *gbp-2* animals. Induction of autophagy is not always detrimental, however. In fact, current evidence indicates that autophagy is activated during dietary restriction, which extends life span [51]. Thus, there appears to be a range in the level of autophagy activation in response to changes in food availability. A moderate level of autophagy within cells appears to be beneficial, possibly providing cells with a variety of metabolic precursors and clearing damaged cellular components. However, high levels of autophagy can be detrimental, contributing to tissue destruction and death.

Another set of factors that may contribute to the decline in muscle mass during aging in worms and other species are micro-RNAs, which modulate gene expression in many different scenarios. Ibenez et al. [52] reported the identification of several micro-RNAs that are age-regulated. Computational prediction of their targets revealed that several age-regulated micro-RNAs in *C. elegans* may disproportionately affect targets expressed in muscles. This led to the hypothesis that muscle aging might reflect changes in target regulation by age-regulated micro-RNAs.

Concluding Remarks

Body composition changes are a major challenge for extending human health span and life span. Increased adiposity and decreased muscle mass can limit the independence of elders and can also predispose to disease. Due to the complex effects of aging on the body, it is difficult to isolate causes and modifying factors of these conditions in clinical studies. Rather, the use of model organisms, including rodents, has been the mainstay for studying these conditions. The utility of invertebrate model organisms, such as *C. elegans* nematodes, has been appreciated in this current era of biogerontology. This chapter highlighted recent studies in *C. elegans* focusing on processes analogous to human adiposity and sarcopenia. These findings may lead to the development of new hypotheses about etiology of these conditions in humans that can be tested using rodent models and clinical studies. In particular, future studies may find therapeutic strategies by targeting neuroendocrine pathways that converge on lipogenic targets. Furthermore, antisarcopenic therapies may be developed that modulate muscle muscarinic or autophagic targets related to those promoting muscle decline during starvation in worms.

References

1 Brenner S: The genetics of *Caenorhabditis elegans*. Genetics 1974;77:71–94.
2 Guarente L, Kenyon C: Genetic pathways that regulate ageing in model organisms. Nature 2000;408: 255–262.
3 Wolkow CA: Identifying factors that promote functional aging in *Caenorhabditis elegans*. Exp Gerontol 2006;41:1001–1006.
4 Hsu AL, Murphy CT, Kenyon C: Regulation of aging and age-related disease by *daf-16* and heat-shock factor. Science 2003;300:1142–1145.
5 Walker GA, Lithgow GJ: Life span extension in *C. elegans* by a molecular chaperone dependent upon insulin-like signals. Aging Cell 2003;2:131–139.
6 Morley JF, Morimoto RI: Regulation of longevity in *Caenorhabditis elegans* by heat shock factor and molecular chaperones. Mol Biol Cell 2004;15:657–664.
7 Tullet JM, Hertweck M, An JH, Baker J, Hwang JY, Liu S, Oliveira RP, Baumeister R, Blackwell TK: Direct inhibition of the longevity-promoting factor *skn-1* by insulin-like signaling in *C. elegans*. Cell 2008;132:1025–1038.
8 McGhee JD: The *C. elegans* intestine. Worm Book 2007:1–36.
9 Ashrafi K: Obesity and the regulation of fat metabolism. Worm Book 2007:1–20.
10 O'Rourke EJ, Soukas AA, Carr CE, Ruvkun G: *C. elegans* major fats are stored in vesicles distinct from lysosome-related organelles. Cell Metab 2009;10: 430–435.
11 Artal-Sanz M, Tavernarakis N: Prohibitin couples diapause signalling to mitochondrial metabolism during ageing in *C. elegans*. Nature 2009;461:793–797.
12 Huang C, Xiong C, Kornfeld K: Measurements of age-related changes of physiological processes that predict life span of *Caenorhabditis elegans*. Proc Natl Acad Sci USA 2004;101:8084–8089.
13 Glenn C, Chow D, David L, Cooke C, Gami M, Iser W, Hanselman K, Goldberg I, Wolkow C: Behavioral deficits during early stages of aging in *Caenorhabditis elegans* result from locomotory deficits possibly linked to muscle frailty. J Gerontol A Biol Sci Med Sci 2005;59A:1251–1260.
14 McKay RM, McKay JP, Avery L, Graff JM: *C. elegans*: a model for exploring the genetics of fat storage. Dev Cell 2003;4:131–142.
15 Brock TJ, Browse J, Watts JL: Genetic regulation of unsaturated fatty acid composition in *C. elegans*. PLoS Genet 2006;2:e108.
16 Kimble J, Sharrock WJ: Tissue-specific synthesis of yolk proteins in *Caenorhabditis elegans*. Dev Biol 1983;96:189–196.

17 Hall DH, Winfrey VP, Blaeuer G, Hoffman LH, Furuta T, Rose KL, Hobert O, Greenstein D: Ultrastructural features of the adult hermaphrodite gonad of *Caenorhabditis elegans*: relations between the germ line and soma. Dev Biol 1999;212:101–123.

18 Narbonne P, Roy R: *Caenorhabditis elegans* dauers need *lkb1*/AMPK to ration lipid reserves and ensure long-term survival. Nature 2009;457:210–214.

19 Watts JL: Fat synthesis and adiposity regulation in *Caenorhabditis elegans*. Trends Endocrinol Metab 2009;20:58–65.

20 Van Gilst MR, Hadjivassiliou H, Jolly A, Yamamoto KR: Nuclear hormone receptor NHR-49 controls fat consumption and fatty acid composition in *C. elegans*. PLoS Biol 2005;3:e53.

21 Van Gilst MR, Hadjivassiliou H, Yamamoto KR: A *Caenorhabditis elegans* nutrient response system partially dependent on nuclear receptor NHR-49. Proc Natl Acad Sci USA 2005;102:13496–13501.

22 Taubert S, Van Gilst MR, Hansen M, Yamamoto KR: A mediator subunit, MDT-15, integrates regulation of fatty acid metabolism by NHR-49-dependent and -independent pathways in *C. elegans*. Genes Dev 2006;20:1137–1149.

23 Kniazeva M, Crawford QT, Seiber M, Wang CY, Han M: Monomethyl branched-chain fatty acids play an essential role in *Caenorhabditis elegans* development. PLoS Biol 2004;2:E257.

24 Ren P, Lim CS, Johnsen R, Albert PS, Pilgrim D, Riddle DL: Control of *C. elegans* larval development by neuronal expression of a TGF-beta homolog. Science 1996;274:1389–1391.

25 Greer ER, Perez CL, Van Gilst MR, Lee BH, Ashrafi K: Neural and molecular dissection of a *C. elegans* sensory circuit that regulates fat and feeding. Cell Metab 2008;8:118–131.

26 Sze JY, Victor M, Loer C, Shi Y, Ruvkun G: Food and metabolic signalling defects in a *Caenorhabditis elegans* serotonin-synthesis mutant. Nature 2000; 403:560–564.

27 Srinivasan S, Sadegh L, Elle IC, Christensen AG, Faergeman NJ, Ashrafi K: Serotonin regulates *C. elegans* fat and feeding through independent molecular mechanisms. Cell Metab 2008;7:533–544.

28 Murakami H, Murakami S: Serotonin receptors antagonistically modulate *Caenorhabditis elegans* longevity. Aging Cell 2007;6:483–488.

29 Murakami H, Bessinger K, Hellmann J, Murakami S: Manipulation of serotonin signal suppresses early phase of behavioral aging in *Caenorhabditis elegans*. Neurobiol Aging 2008;29:1093–1100.

30 Chen S, Whetstine JR, Ghosh S, Hanover JA, Gali RR, Grosu P, Shi Y: The conserved NAD(H)-dependent corepressor CTBP-1 regulates *Caenorhabditis elegans* life span. Proc Natl Acad Sci USA 2009; 106:1496–1501.

31 Budovskaya YV, Wu K, Southworth LK, Jiang M, Tedesco P, Johnson TE, Kim SK: An *elt-3/elt-5/elt-6* GATA transcription circuit guides aging in *C. elegans*. Cell 2008;134:291–303.

32 Libina N, Berman JR, Kenyon C: Tissue-specific activities of *C. elegans daf-16* in the regulation of life span. Cell 2003;115:489–502.

33 Honda Y, Honda S: The *daf-2* gene network for longevity regulates oxidative stress resistance and Mn-superoxide dismutase gene expression in *Caenorhabditis elegans*. FASEB J 1999;13:1385–1393.

34 Murphy CT, McCarroll SA, Bargmann CI, Fraser A, Kamath RS, Ahringer J, Li H, Kenyon C: Genes that act downstream of DAF-16 to influence the life span of *Caenorhabditis elegans*. Nature 2003;424:277–283.

35 Kenyon C, Chang J, Gensch E, Rudner A, Tabtiang R: A *C. elegans* mutant that lives twice as long as wild type. Nature 1993;366:461–464.

36 Kimura KD, Tissenbaum HA, Liu Y, Ruvkun G: *daf-2*, an insulin receptor-like gene that regulates longevity and diapause in *Caenorhabditis elegans*. Science 1997;277:942–946.

37 Wang MC, O'Rourke EJ, Ruvkun G: Fat metabolism links germline stem cells and longevity in *C. elegans*. Science 2008;322:957–960.

38 Aiken J, Bua E, Cao Z, Lopez M, Wanagat J, McKenzie D, McKiernan S: Mitochondrial DNA deletion mutations and sarcopenia. Ann NY Acad Sci 2002;959:412–423.

39 Edstrom E, Altun M, Bergman E, Johnson H, Kullberg S, Ramirez-Leon V, Ulfhake B: Factors contributing to neuromuscular impairment and sarcopenia during aging. Physiol Behav 2007;92: 129–135.

40 Garigan D, Hsu A-L, Graser AG, Kamath RS, Ahringer J, Kenyon C: Genetic analysis of tissue aging in *Caenorhabditis elegans*: a role for heat-shock factor and bacterial proliferation. Genetics 2002;161:1101–1112.

41 Herndon LA, Schmeissner PJ, Dudaronek JM, Brown PA, Listner KM, Sakano Y, Paupard MC, Hall DH, Driscoll M: Stochastic and genetic factors influence tissue-specific decline in ageing *C. elegans*. Nature 2002;419:808–814.

42 Chow DK, Glenn CF, Johnston JL, Goldberg IG, Wolkow CA: Sarcopenia in the *Caenorhabditis elegans* pharynx correlates with muscle contraction rate over life span. Exp Gerontol 2006;41:252–260.

43 Larsen PL: Aging and resistance to oxidative damage in *Caenorhabditis elegans*. Proc Natl Acad Sci USA 1993;90:8905–8909.

44 Avery L: The genetics of feeding in *Caenorhabditis elegans*. Genetics 1993;133:897–917.

45 Johnston J, Iser WB, Chow DK, Goldberg IG, Wolkow CA: Quantitative image analysis reveals distinct structural transitions during aging in *Caenorhabditis elegans* tissues. PLoS One 2008;3:e2821.

46 You YJ, Kim J, Cobb M, Avery L: Starvation activates map kinase through the muscarinic acetylcholine pathway in *Caenorhabditis elegans* pharynx. Cell Metab 2006;3:237–245.

47 Kang C, Avery L: Systemic regulation of autophagy in *Caenorhabditis elegans*. Autophagy 2009;5:565–566.

48 Cuervo AM, Bergamini E, Brunk UT, Droge W, French M, Terman A: Autophagy and aging: the importance of maintaining 'Clean' cells. Autophagy 2005;1:131–140.

49 Reynolds THt, Krajewski KM, Larkin LM, Reid P, Halter JB, Supiano MA, Dengel DR: Effect of age on skeletal muscle proteolysis in extensor digitorum longus muscles of B6C3F1 mice. J Gerontol A Biol Sci Med Sci 2002;57:B198–B201.

50 Kang C, You YJ, Avery L: Dual roles of autophagy in the survival of *Caenorhabditis elegans* during starvation. Genes Dev 2007;21:2161–2171.

51 Hansen M, Chandra A, Mitic LL, Onken B, Driscoll M, Kenyon C: A role for autophagy in the extension of life span by dietary restriction in *C. elegans*. PLoS Genet 2008;4:e24.

52 Ibanez-Ventoso C, Yang M, Guo S, Robins H, Padgett RW, Driscoll M: Modulated microrna expression during adult life span in *Caenorhabditis elegans*. Aging Cell 2006;5:235–246.

Catherine A. Wolkow
NIA IRP, NIH, Biomedical Research Center, Suite 05C222
251 Bayview Blvd.
Baltimore, MD 21224 (USA)
Tel. +1 410 558 8566, Fax +1 410 558 8323, E-Mail wolkowca@mail.nih.gov

Mobbs CV, Hof PR (eds): Body Composition and Aging.
Interdiscipl Top Gerontol. Basel, Karger, 2010, vol 37, pp 94–114

Sarcopenia: Prevalence, Mechanisms, and Functional Consequences

Michael J. Berger[a] · Timothy J. Doherty[a,b]

[a]School of Kinesiology, and Departments of [a]Clinical Neurological Sciences and [b]Physical Medicine and Rehabilitation, Schulich School of Medicine, University of Western Ontario, London, Ont., Canada

Abstract

Aging is associated with significant decline in neuromuscular function and performance. Sarcopenia, often defined as age-related loss of muscle mass, strength, and functional decline, is the most characteristic feature of age-related changes in the neuromuscular system. Strength decline in upper and lower limb muscles is typically 20–40% by the 7th decade and greater in older adults. This is accompanied by similar losses of limb muscle cross-sectional area. Whole body or appendicular muscle mass determination has become the method of choice for defining sarcopenia. Large population studies have reported that sarcopenia affects over 20% of 60- to 70-year-olds, and approaches 50% in those over 75 years. While loss of muscle mass explains a significant component of weakness, other factors are emerging as important contributors. In particular changes at the level of the motor neuron and motor unit are discussed. Muscle power has emerged as an important indicator of function in older adults, and we discuss knee osteoarthritis as a model of accelerated limb sarcopenia.

It is well established that the human aging process is associated with a significant decline in neuromuscular function and performance [1–3]. Characteristic of this decline is the inevitable reduction in skeletal muscle mass and associated loss of strength, so-called *Sarcopenia*, that occurs even in the absence of disease specifically affecting the neuromuscular system. Rosenberg [4] first coined the term sarcopenia in 1989, literally meaning poverty of flesh, to describe age-associated loss of skeletal muscle mass. Subsequently, the definition has been broadened to imply loss of muscle mass, strength, muscle contractile quality and functional decline. To date, there is no universally accepted definition of sarcopenia, which, in part, is related to the absence of well-accepted standardized methodology or criteria to establish its presence in a given population. Nevertheless, recent expert opinion [5] has suggested that the definition of sarcopenia encompass age-related alteration in muscle mass, strength, and function and this broader definition will be adopted in this chapter.

This review will examine sarcopenia in the context of age-related losses of muscle mass and strength, its epidemiology, and potential etiologic factors. In particular, the role of motor neuron loss and its impact on muscle fiber numbers and muscle mass will be examined. Age-related reduction in muscle power has recently emerged as a stronger influence on function than strength, and this will be reviewed in greater detail. Finally, knee osteoarthritis as a potential model of 'accelerated sarcopenia' in the lower limb is discussed.

Age-Related Decline in Strength

Loss of skeletal muscle strength is a commonly recognized consequence of aging. Age-related decline in strength has been demonstrated in multiple cross-sectional studies of limb muscles tested under isometric and dynamic (usually isokinetic) conditions, most often comparing groups of healthy young subjects with, middle-aged, and older men and women [1, 6, 7]. The knee extensors, because of their functional importance, ease of testing, and presence of comparative biopsy data, have been most frequently examined.

Multiple studies have compared knee extensor strength in groups of young with healthy older subjects in their seventh and eighth decades. The average reported age-related decreases in strength are on the order of 20–40% [8–12]. Even greater losses (50% or more) have been reported for those in their 9th decade and beyond [11–13]. In general, similar declines in strength have been reported for proximal and distal limb muscles including the ankle plantar and dorsiflexors, elbow flexors and extensors, and hand grip [14–20]. Relative losses appear similar for men and women; however, since men typically start from higher baseline values, their absolute losses of strength are greater. In concert with age-related slowing of electrically evoked muscle contractile properties [16, 21–26], some reports have shown more significant losses of strength with isokinetic testing at higher angular velocities [8, 15, 27]. One important exception to these observations is the consistent finding of relative preservation of strength under eccentric testing conditions [28, 29]. It has been postulated that this preservation of high velocity strength may be related to slower contractile properties, slower cross-bridge cycling, and increased or altered connective tissue content and muscle stiffness in older adults.

While it has been well established that short duration resistance training interventions can partially reverse losses of strength in even the very old [6, 7, 30, 31], the extent to which life-long activity patterns and training can prevent age-related declines in strength has not been prospectively examined. However, Klitgaard et al. [32], in a cross-sectional study, compared elderly men (mean age 69 years) who had either trained with running, swimming or strength training regularly for between 12 and 17 years with young and elderly sedentary controls. In comparison to the young controls, they reported strength declines in the sedentary elderly group for maximal isometric torque for the knee extensors (44%) and elbow flexors (32%). However, while the older swimmers and runners exhibited similar declines in strength to the sedentary elderly, the strength-trained elderly men had similar maximal isometric strength and muscle cross-sectional areas as the young controls. These results are clearly limited by the cross-sectional study design and potential selection bias, but do provide evidence that, at least in a selected population, strength losses with aging may be attenuated by long durations of resistance exercise.

Thus, it appears that healthy men and women in their seventh and eighth decades exhibit, on average, 20–40% less strength in comparison to their younger counterparts. These losses are even greater (50% or more) for the very old. In general, similar losses are present for proximal and distal muscles in the upper and lower extremities and men and women experience similar losses on a relative basis. Longitudinal studies, with some exceptions, have reported somewhat greater losses of strength over time (1–3%/year) in comparison to cross-sectional studies. The majority of these studies, however, have examined older populations over a limited duration of follow-up, and thus may not predict the rate of decline in young or middle-aged populations.

Age-Related Loss of Skeletal Muscle Mass

Age-related declines in strength are directly impacted by, and correlated with, losses of skeletal muscle mass. It has been demonstrated that total muscle cross-sectional area decreases by about 40% between the ages of 20 and 80 [3, 6, 7]. Cross-sectional areas have been determined for various limb muscle groups using ultrasound, computed tomographic scanning, magnetic resonance imaging, and direct measurement of whole muscle cross sections from cadaveric specimens. For example, Young et al. [11, 12] using ultrasonographic imaging, reported 25–35% reductions in the cross-sectional areas of the quadriceps muscles in older men and women as compared to young controls. Computed tomographic scanning has shown similar results for the quadriceps muscle [33], the biceps brachii [34] and triceps brachii [34] in men. Additionally, highlighting the inaccuracy of simple anthropometric measures of limb circumference, Rice et al. [34] found 27, 45, and 81% more nonmuscle tissue (fat and connective tissue) for the arm flexors, arm extensors and plantar flexors, respectively. Similarly, Overend et al. [33] reported increases in nonmuscle tissue of 59% for the quadriceps and 127% for the hamstrings. Cross-sectional area measures taken directly from whole muscle obtained postmortem showed similar average reductions of 40% between 20 and 80 years of age [35]. The average reduction was 10% at 50 years and accelerated thereafter.

More recent studies have attempted to define sarcopenia through quantifying whole body muscle mass or appendicular skeletal muscle mass (ASM) by a variety of methods, with magnetic resonance imaging (MRI) and dual energy X-ray absorptiometry (DEXA) emerging as the two most well accepted methods. Indeed, current expert opinion suggests that determination of whole body, or appendicular muscle mass comprises an important aspect of defining sarcopenia [5]. Baumgartner et al. [36], in the New Mexico Elder Survey, measured appendicular muscle mass by DEXA in 883 randomly selected elderly Hispanic and white men and women. Sarcopenia was defined as loss of muscle mass greater than 2 SD below the mean for young healthy controls. The prevalence of sarcopenia ranged from 13 to 24% in persons aged 65–70 years and was over 50% for those older than 80 years. In this study, the prevalence was higher for men over age 75 (58%) than for women (45%). In a similar study, the prevalence based on total skeletal mass determined by DEXA was 10% for men and 8% for women between 60 and 69 years, and 40% and 18%, respectively, for men and women over 80 years [37]. As these prevalence rates are relative to gender-specific younger control populations they suggest greater declines in muscle mass for men than women. Iannuzzi-Sucich et al. [38] used DEXA to quantify appendicular skeletal muscle mass in 195 women aged 64–93 years and 142 men aged 64–92 years. They defined sarcopenia as 2 SDs below the muscle mass/height $(m)^2$ for young controls. The overall prevalence of sarcopenia so defined was 22.6% in women and 26.8% in men. These values climbed to 31 and 45%, respectively, for women and men over 80. Similarly, Janssen et al. [39] used whole body MRI to examine skeletal muscle mass and distribution in a large cohort of 468 men and women from 18 to 88 years of age. They observed a decline in whole body muscle mass beginning in the third decade; however, this did not become substantial until the end of the 5th decade. An important finding of this study was that the loss of muscle mass with aging was greater in the lower body in men and women. This finding may reflect decreased activity or altered patterns of activity of the lower extremity muscles with aging and has important implications for functional mobility and disability. In summary, the prevalence of sarcopenia is partially dependent on the population studied, the measurement technique employed and the operational definition used. However, in general, sarcopenia affects at least 20% of men and women in their 7th decade and often approaches 50% or greater for those over 75 years.

While sarcopenia most literally refers to loss of skeletal muscle mass and has been extended to imply loss of strength, clearly functional ability is of the utmost importance to elderly men and women in terms of independence and quality of life. It would seem intuitive that a relationship should exist between muscle mass, strength and the ability to carry out functional tasks. This was evident in the New Mexico Elder study where sarcopenic women had 3.6-fold higher rates of disability and men had 4.1-fold higher rates in comparison to those with greater muscle mass [36]. The use of assistive walking aids and falls were also higher in these subjects. More recently, Janssen et al. [40] reported a higher prevalence of more severe sarcopenia in women as compared to men over 60 years. Functional impairment and disability were two times greater in older men and three times greater in older women.

Alternatively, Visser and colleagues [41] examined 3,075 adults over 70 years and reported that whole body and lower limb strength, as opposed to leg muscle mass (based on DEXA), were independent predictors of lower extremity performance. Similar results were reported by their group from a cross-sectional analysis of 449 men and women 65 years of age and older whereby low strength (grip strength), but not low muscle mass (from DEXA), were associated with poor lower extremity performance as measured by gait speed and chair rise. The results of a 5-year longitudinal study of thigh muscle mass, intramuscular fat infiltration and strength in men and women over 70 years of age reported 2- to 5-fold greater losses of muscle strength (isokinetic torque) than muscle mass. This was accompanied by significant increases in intramuscular fat, which was thought to contribute to decreased muscle contractile quality. Others have reported similar greater losses of strength over time, with one longitudinal study reporting that the decline in strength over time was 60% greater than predicted from cross-sectional analysis and that age-related loss of muscle mass accounted for only 5% of the variance in the loss of strength [42]. As a result of these and other observations, it has become clear that loss of muscle mass is only partially responsible for age-related loss of strength and functional mobility and other factors including central drive, muscle contractile quality, and altered excitation-contraction coupling to name a few are implicated [43].

Mechanisms Underlying Sarcopenia

Multiple, inter-related factors contribute to the development and progression of sarcopenia [44]. These factors, no doubt, contribute in varying degrees to the age-related losses of muscle mass, strength, muscle quality and the degree of functional impairment and reserve present in older men and women. It is also probable that certain underlying mechanisms are of greater influence than others when considering any specific age group, gender or association with co-morbid states. Among others, altered endocrine function (growth hormone, insulin, estrogen, testosterone), increased levels of proinflammatory cytokines (e.g. IL-6, TNF), mitochondrial dysfunction, cellular apoptosis, and inadequate nutrition (particularly dietary protein) have all been implicated as potential contributing factors to loss of muscle mass, strength and contractile quality. These potential etiologic factors have recently been reviewed in detail [44]. In addition to these factors, we and others have shown in multiple studies that there is a strong neurogenic component underlying age-related loss of muscle mass and strength. This has been established primarily through the use of electrophysiological techniques and specifically 'motor unit number estimation' or MUNE [45, 46]. MUNE techniques estimate the numbers of functioning MUs in a muscle group by dividing a size parameter of the electrically evoked compound muscle action potential (the muscle's maximal output) by the mean size of its constituent motor unit potentials. These observations related to aging and sarcopenia are summarized below.

Regardless of any potential age-related impairment in contractile quality, loss of muscle mass remains as a significant contributing factor to strength decline and associated disability in older men and women. Histologic data, predominately from needle biopsy sampling, has provided some insight into the cause of the age related atrophy. The majority of these data are from the vastus lateralis muscle and the overall findings are reasonably consistent. That is, the average type II fiber size is diminished with age while the size of type I fibers is much less affected [3, 7, 8, 35, 47–50]. While reductions in type II area range from 20 to 50%, type I fiber area losses range from 1 to 25%. The variability noted in these studies relates to sampling variability, potential sampling bias with muscle biopsy, and the undoubted inherent variability in both the older and younger control populations.

The above reductions in fiber size, however, are typically moderate in comparison with the reductions in muscle mass, and therefore reductions in muscle fiber number have been proposed. In a landmark study, Lexell's group [35, 48] using whole muscle cross-sections from the vastus lateralis muscle obtained postmortem, reported that by the ninth decade approximately 50% fewer type I and type II fibers were present in comparison to muscles from 20-year-olds. The observation that similar losses of muscle fibers occurred for both type I and II fibers stood in contrast to earlier work from samples obtained with muscle biopsy [8, 51]. Further analysis determined that the cross-sectional area of the vastus lateralis was mainly determined by the total number of fibers and, to a lesser extent, by the size or number of type II fibers [35, 48, 49]. There was no way, however, based upon these analyses to determine the etiology of the fiber loss.

Further to this, in concert with type II atrophy, there is histochemical evidence of fiber type grouping, fiber atrophy, and increased co-expression of myosin heavy chain isoforms in the same fiber, thought consistent with a progressive denervation and reinnervation process secondary to a chronic neuropathic process [52–54]. Given these findings and the previously noted losses of muscle fibers, it has been suggested that alpha motor neuron loss may be largely responsible for age-related loss of muscle mass [23, 46, 50, 55, 56].

Is there any support for such a hypothesis? Electrophysiological studies using either macro electromyographic techniques [57] or motor unit number estimation techniques [45, 46] have demonstrated substantive losses of whole functioning motor units in proximal and distal muscles in the upper and lower extremities [46, 58, 59]. These reported losses are on the order of 50% for the thenar, hypothenar, and biceps/brachialis muscle groups [23, 55, 56, 60, 61] and are consistent with anatomic data that have demonstrated losses of anterior horn cells and ventral root fibers with aging [62–65]. These findings, taken together with muscle morphologic changes consistent with a chronic neuropathic process point toward age-associated losses of motor neurons/motor nerves as an important contributing factor to reduced muscle fiber number and muscle mass [3, 66, 67]. No longitudinal studies have examined this process, but cross-sectional studies would suggest that motor neurons or motor unit numbers are well maintained until the seventh decade and then begin to decline precipitously thereafter [21, 68, 69].

Our group recently examined the impact of MU loss as a potential contributing factor to progressive or accelerated strength loss in the ankle dorsiflexors of very old men [70]. Voluntary strength, evoked twitch strength, maximum compound muscle action potential (CMAP) size, and MUNE (the number of functioning motor units) were determined in a cross-sectional study from a group of young (mean age 27 years), old (66 years) and very old (82 years) men. The number of MUs in the tibialis anterior muscle was reduced in the older men (91 MUs) in comparison to the young (150 MUs); further reductions still were observed in the very old (59 MUs). Despite significant MU losses, twitch strength, CMAP size and MVC were maintained in the old group – likely as a result of effective collateral reinnervation of denervated muscle fibers [3]. However, in the very old, collateral reinnervation was unable to maintain pace with the extent of motor unit loss resulting in substantive loss of both voluntary and evoked strength. This raises the concept that the neuromuscular system may be able adapt to loss of motor units until a threshold is reached, after which progressive losses of muscle mass and strength occur – this, in part, may explain the rapid decline in strength that occurs in the very old leading to mobility disability and subsequent frailty. Interestingly, there appears to be variability in these findings in different muscle groups, as in a second series of experiments, we did not observe the same extent of age-related losses of motor units in the soleus [71]. Perhaps this relates to differences in the underlying fiber type/MU type (soleus is predominantly composed of type 1 muscles fibers), altered patterns of use or function of a given muscle group.

The extent to which chronic physical activity might affect motor neuron function is intriguing. To that end, we recently examined the numbers of motor units in a group of masters distance runners (mean age 65 years). While their MVC strength was lower than young controls and sedentary older adults (a common observation for endurance athletes likely related to low body mass), the estimated number of motor units in the tibialis anterior muscle of the masters runners was similar (140 MUs) to young control data [72]. These preliminary results require further substantiation, but provide support for physical activity as a potential protective factor in maintaining the health and function of the motor neuron, possibly as a result of reduced oxidative stress.

Age-Related Changes in Power

Age-related reduction in muscle power has received far less attention than strength. Muscle power is the product of force and velocity of movement and may be measured during dynamic and isokinetic (constant velocity) contractions. There is evidence to suggest that power declines earlier in the lifespan and more precipitously than force [73, 74]. Additionally, power may be a more robust predictor of functional ability in the elderly [75]. Our discussion of muscle power will consider: (1) reports of age-related power deficits and altered force-velocity relationships in the elderly with regard to age, muscle type and contraction mode, as well as purported mechanisms; (2) the relationship between muscle power and function, and (3) the various strength and power training exercise programs designed to attenuate age-related losses in power.

Age-Related Change in Muscle Power and the Force-Velocity Relationship

The ultimate design constraints imposed on a muscle during movement are defined by its force-velocity relationship [76]. Muscle may perform at or below this relationship, but never above it. Also, while maximal isometric strength and unloaded shortening velocity are theoretical constraints, skeletal muscle usually performs along a continuum defined by the middle of the force-velocity relationship during usual daily activities. Aging is accompanied by a leftward shift in the force-velocity curve and reduced curvature of the relationship in the concentric domain. As a result, the power-velocity curve is shifted leftward and down so that: (1) peak power (defined by the point at which both force and velocity are optimized) is reduced, and (2) peak power occurs at a slower optimal velocity [77]. It should be noted that the characteristics of the force-velocity relationship are different in the eccentric domain. It is well established that eccentric strength and power are relatively well preserved in elderly subjects; however, the mechanisms for this phenomenon remain to be fully elucidated. The reader is referred to a recent review by Raj et al. [77] for a more detailed discussion.

Although the evidence is limited by the cross-sectional design employed by most studies, it appears that the age-associated power decrement proceeds at a faster rate and is of greater magnitude than isometric strength loss. A mean deficit of 3.5%/year for knee extensor power versus 1–2%/year for isometric strength between ages 65 and 89 has been reported [73]. Furthermore, peak power deficits in elderly exceed isometric strength deficits [78–82]. For example, Valour et al. [80] reported a deficit of 45% for mean isokinetic elbow flexor power for elderly subjects (age range 61–78 years) versus young subjects (age range 18–27 years) and an isometric elbow flexor torque deficit of 28%. Similar results have been found in the lower limb for knee extensors [78, 79] and dorsiflexors [79, 82].

Muscle power in vivo has been measured with isokinetic [79, 83, 84], isotonic [80–82, 85] or explosive multi-joint movements [86, 87]. Power can also be inferred from studies using isokinetic contractions to model torque-velocity relationship wherein power is not measured directly [88, 89]. Regardless of the inter-study variability in muscle studied, contraction mode or age range, when a cross-section of older subjects is compared to younger subjects power is invariably reduced. Furthermore, the magnitude of power deficit has been shown to increase at higher contraction velocities, suggesting that activities requiring rapid development of force become more difficult with age [79, 80, 84, 88, 90]. As well, intermuscle differences in power deficits may be present with some muscle groups more susceptible than others. For example, Lanza et al. [79] reported that isokinetic power deficits in elderly subjects were more pronounced for knee extensors than for dorsiflexors. This observation was attributed to the fact that each muscle group may be differentially susceptible to the mechanisms responsible for power reduction because of their respective muscle fiber and motor unit compositions (i.e. knee extensors are more susceptible than dorsiflexors to the age-related reduction in number and proportion of fibers expressing type IIa and IIx myosin ATPase).

With regard to the mechanisms of age-related loss in power, it appears that adaptations other than sarcopenia alone are responsible. While the loss of muscle mass, particularly of type II muscle fibers, could explain power deficits in part, when the age-related reduction in muscle mass is controlled for by normalizing power to a measure of muscle cross-sectional area (i.e. specific power), deficits in power still exist between young and old subjects [82]. Studies of whole muscle power during jumping tasks [87, 91] and isotonic dorsiflexion contractions [82] suggest that reductions in the velocity component of power are primarily responsible. Furthermore, some studies report reductions in specific power without reduced specific tension, which further implicates those muscle properties responsible for velocity [82, 84, 85]. Potential causes of velocity reduction include reduced proportion of type II muscle fibers [92], reduced number of sarcomeres in series [84], and changes in functional myosin of the heavy chain (MHC) isoform [93]. At the single fiber level, the most important determinant of power is MHC isoform [94]. There is evidence to suggest ATPase enzyme activity within a given muscle fiber type is altered; however studies of single muscle fiber contractile properties in the knee extensors are equivocal. Some investigators report an age-related slowing in single fiber maximal unloaded shortening velocity for type I and IIa fibers, which would explain the decrease in specific power shown at the whole muscle level [90, 95–98]. Trappe et al. [99] measured maximal isometric force, unloaded shortening velocity, power and cross-sectional area of single muscle fibers and reported similar specific tension and specific power between young and old type I and IIa muscle fibers. Small sample size, differences in subject populations and lack of control over physical activity were cited to explain the discrepant findings [99]. At present, it is unknown whether quantitative changes in whole muscle (i.e. reduced muscle mass, reduced proportion of type II fibers) or qualitative changes in individual fibers, is the primary cause of age-related velocity reduction.

Relationship between Muscle Power and Mobility Performance

Scores attained on performance tests related to mobility are important predictors of outcomes such as falling, institutionalization and mortality [100–102]. Consequently, it has been of considerable interest to identify clinically relevant impairments associated with reduced mobility performance. The relationship between lower-limb strength impairment and function has been well established [103–105]; however, some studies report lower strength of association for low-intensity performance measures such as habitual gait velocity [106, 107]. Early studies also established a relationship between muscle power impairments and functional mobility tests such as stair climbing, rising from a chair and timed walk [108, 109]. Recent studies have standardized the definition of mobility performance by using composite measures such as the Short Physical Performance Battery (SPPB) [101]. The evidence suggests that lower-limb muscle power is a more robust predictor of performance than strength and the relationship often improves when controlling for body mass, age and sex [110–114]. Bean et al. [110] reported that bilateral and unilateral leg extensor peak power measured in older adults (age range 65–83) with mobility limitations explained a greater proportion of the variance in performance on the SPPB and other mobility tests, rather than strength. This has also been confirmed for some performance measures in the plantar flexors and dorsiflexors and in the upper limb [115, 116]. In another study by Bean's group, power, strength and function were transformed into categorical variables (i.e. high power group vs. low power group). Those with the lowest levels of strength and power had the greatest likelihood of mobility limitation; however, the odds ratios were 2- to 3-fold higher for muscle power [117]. Many activities of daily living are velocity-dependent, requiring dynamic contractions, thus the emphasis on power over strength is intuitive.

A reduction in the velocity component of muscle power may be an independent predictor of function. Mayson et al. [118] reported that leg extensor velocity at a contraction intensity of 40% of 1 repetition maximum for a squat exercise, was strongly associated with composite measures of balance. In a follow-up study, when velocity and strength were transformed into categorical variables, a stronger relationship to SPPB score was reported for velocity [75].

Due to the influence of velocity on performance, the relationship between muscle power and function may be activity-dependent [106, 111, 113, 119]. In other words, it could be that only some performance measures show associations with power. Puthoff and Neilsen [119] reported that leg extensor power measured at a higher relative intensity better predicts performance on a chair rise test, while peak power (measured at a lower intensity) was a stronger predictor of speed-related activities such as a 6-min walk test. The implication being that power must be improved at multiple contraction intensities to enhance both high- and low-intensity functional activities.

A linear relationship between power and function reported in some studies suggests that any improvement in muscle power should be associated with improvements in function. The results of other studies suggest a strengthening of the predictive value of power to performance when the relationship is described with a curvilinear function [113, 117]. A curvilinear relationship suggests that there is a threshold whereby improvements in power no longer correspond to improvements in function. Consequently, the lowest functioning patients would likely be more sensitive to small changes in power. A similar relationship has been reported for isometric strength [120]. Those studies reporting a linear relationship between power and function may have a selection-bias towards the least powerful subjects. Therefore, those individuals that are weakest may respond best to improvements in muscle power.

Training

The age-related power deficit is a clinically relevant impairment because it is associated with reduced function and it is amenable to treatment with both traditional strength training (ST) and power training (PT) exercise interventions. Investigators have long known that strength can be improved with ST programs (i.e. concentric and eccentric phases of contraction lasting ≥2 s), even in nonagenarians [30]. With the recent interest in measures of muscle power as a robust predictor of function, a series of randomized trials have investigated whether PT programs emphasizing velocity of contraction are more effective for improving strength and power in older subjects. In general, PT interventions have employed either leg press or knee extensor exercises, are of shorter duration (8–16 weeks), with a frequency of 2–3 visits per week and 3–4 sets/8–12 repetitions per exercise performed at 20–80% of 1 repetition maximum (1RM). The duration of the concentric phase of contraction in PT differentiates it from ST, as PT exercises are performed 'as fast as possible'. PT programs are shown to be effective for improving strength, power and function in older adults [121–123]. Improvement in the force component of power is thought to be the mechanism by which PT improves power, as velocity of contraction may actually be reduced during the course of such an intervention [124]. The optimal contractile intensity for a PT program has yet to be determined, with studies reporting no benefit for one particular intensity over another [122, 124, 125], although training with low weight and high velocity contractions may improve performance on some functional tests, such as balance [122].

Randomized controlled trials comparing ST versus PT are equivocal. Generally, both interventions yielded positive benefits versus a control group, although some studies report enhanced benefits for the PT group for a variety of outcome measures [126–130], while others do not [131, 132]. When strength is used as a primary outcome measure, it is evident that both interventions confer similar benefits [127, 129–131]. However, when power is used both Henwood et al. [131] and Reid et al. [132] report no benefit for PT over ST. The reasons for this discrepancy are unclear and may depend on differences in study population, whether the study was adequately statistically powered, duration of intervention and method for assessing muscle power. Despite the negative results, the effect of each intervention on secondary outcome measures in these studies pointed towards added benefit for PT over ST. For example, Henwood et al. [131] found similar gains in strength and power, but the PT group performed 20% less work during exercise. Reid et al. [132] found that leg press specific power improvement was greater in the PT group. Therefore, with respect to some physiological outcome measures, it appears that PT is more effective. Further to these findings, others measured the effect of ST and PT on function and disability. Studies using the SPPB as a measure of function showed similar improvements between ST and PT [126, 129], while those using different measures of function and self-reported quality of life measures reported added benefit for PT [127, 128]. Again, there is indirect evidence from the negative studies in favor of PT. Bottaro et al. [127] used the SPPB as a secondary outcome measure and therefore study power was inadequate to detect between-group differences. Also, Bean et al. [126] sub-classified participants into high and low velocity groups according to baseline measures and reported that SPPB improved with PT and not ST in the low velocity group. Further adequately powered studies are required to determine whether there is a clear benefit for PT over ST.

Muscle Weakness and Osteoarthritis of the Knee

Osteoarthritis (OA) is one the most common and chronic conditions of aging. Data from the Framingham study reveal that disability caused by OA equals heart disease and is greater than other common diseases of aging such as diabetes and COPD [133]. The knee is the most frequently affected joint in the lower limb and contributes to significant mobility impairment. Up to 65% of people over age 65 years old have radiographic evidence of knee OA [134]. Traditionally, knee OA is characterized by pathological changes such as attrition of the articular hyaline cartilage, marginal osteophyte formation, and subchondral sclerosis and cyst formation. However, a more recent model asserts that knee OA affects all joint structures including cartilage, bone, ligaments, synovium, menisci and muscle [135]. Quadriceps muscle weakness in particular has strong association with mobility limitations and pain in knee OA and may be the affected structure most amenable to conservative treatment [136, 137]. Consequently, research investigating the etiology, mechanisms and treatments for muscle weakness in knee OA is of interest.

The quadriceps muscles of patients with knee OA are weaker than their healthy age-matched counterparts or a nondiseased contralateral limb. Quadriceps muscle strength in knee OA is reduced between 20 and 40% for isometric [138–142] and isokinetic [143–146] contractions. The consequences of reduced quadriceps strength are twofold: (1) the quadriceps may act to protect the knee joint from harmful loading forces, thereby preventing disease incidence and progression, and (2) quadriceps muscle impairment leads to disability.

There is indirect evidence to suggest the breakdown of neuromuscular protective mechanisms around the knee joint (e.g. quadriceps muscle weakness) increases the risk for development of knee OA. In two cross-sectional studies, quadriceps weakness was significantly correlated with radiographic knee OA even in those who did not yet describe symptoms of knee pain [134, 145]. It is important, however to remember that correlation does not imply causation. In a longitudinal prospective cohort study, Slemenda et al. [147] followed healthy women for a mean of 31.3 months and reported that quadriceps muscle strength in women who developed radiographic knee OA was 18% lower than those who did not. Further evidence comes from Becker et al. who reported reduced quadriceps strength in a group of patients who had undergone medial meniscectomy (presumably a 'preosteoarthritic' state), compared to healthy controls [148].

Biomechanical analysis of gait may provide a plausible mechanism to explain how quadriceps muscle weakness causes incident and worsening disease. Eccentric contraction of the quadriceps muscle prior to and during heel strike provides deceleration of the lower limb and shock absorption from the loading impulses generated at heel-strike during the gait cycle [149]. A few studies report increases in loading forces in models of weak quadriceps, but these findings have not been extended to models of knee OA [150, 151]. A more likely scenario involves a complex interplay of neural gait strategy, knee joint alignment and quadriceps function causing abnormal loading forces on a susceptible knee joint [152, 153]. In animal models, quadriceps weakness induced by botulinum injection actually caused reduced loading forces across the knee joint, but led to degenerative changes in patellofemoral cartilage [154, 155]. Despite conflicting evidence as to the exact mechanism, altered quadriceps muscle function is implicated in disease initiation.

Perhaps more relevant is the relationship between muscle weakness and disability in knee OA. Radiographic disease severity is usually a poor predictor of clinically relevant outcomes such as pain and disability [156, 157]. Two large population studies of community dwelling subjects reported that muscle weakness is a stronger predictor of disability and pain than radiographic disease severity [136, 137]. A longitudinal study following women age 70–79 over 72 months reported that reduced knee extensor strength was an independent predictor of the transition between normal mobility and mobility difficulty [158]. Thus, muscle weakness is detrimental to function in knee OA just as in healthy aging, regardless of radiographic disease severity.

The traditional model of muscle weakness in knee OA is one whereby chronic disuse due to knee pain precipitates muscle atrophy, resulting in further disuse. There are a few studies providing direct evidence that muscle atrophy is a primary mechanism of quadriceps muscle weakness. Ikeda et al. [159] reported a 12% reduction in quadriceps muscle mass in women who displayed evidence of incident radiographic knee OA compared to those who did not. Subsequently, Peterson et al. [160] reported that muscle atrophy explained 27% of the variance in isometric quadriceps strength in those with end-stage radiographic knee OA. It is possible that quadriceps muscle weakness in knee OA represents a process of 'accelerated aging' whereby the mechanisms of sarcopenia (e.g. muscle fiber loss, increased proportion of the type I muscle fibers and selective type II fiber atrophy) are occurring at a faster rate. In support of this theory, a recent study by Ling et al. [142] assessed motor unit properties during voluntary contractions in patients with radiographic knee OA. During low-intensity contractions, the mean amplitude of the surface motor unit action potential sampled with needle electromyography was higher for knee OA patients than for controls. This finding is consistent with the motor unit denervation-collateral reinnervation process documented in healthy aging, which accounts for larger motor units being recruited at low-intensity contractions [7]. Further study of muscle contractile properties and biochemical features is required to determine the causes of muscle atrophy.

Beyond disuse atrophy, there is ample evidence supporting impairment in neuromuscular function as a cause of reduced strength. Numerous studies report a reduction in the ability of the central nervous system to recruit and maximally rate code all available motor units during maximal voluntary contraction (i.e. voluntary activation deficits) [138–140, 148, 160–162]. Type Ia afferents from the knee joint may be inhibited by knee pain or effusion, leading to reduced alpha-motor neuron excitability during voluntary contractions, a mechanism termed 'arthrogenous muscle inhibition' [139]. Other studies have shown no difference in voluntary activation between knee OA and healthy subjects, in particular in younger subjects with less severe disease [141, 163]. It is possible that disease severity determines the ability to maximally activate the quadriceps, although this has yet to be examined.

Another proposed neuromuscular mechanism for reduced strength is an increase in co-contraction of the antagonist muscles, perhaps in an attempt to stabilize an abnormal knee joint [164]. At this point, only one study has investigated co-contraction as a mechanism for reduced quadriceps muscle strength in knee OA with negative results [163]. Until the mechanisms of muscle weakness in knee OA are accurately characterized, the design of appropriate rehabilitation strategies is challenging.

Strength training for improvement of quadriceps muscle strength is acceptable conservative management for treatment of knee OA [165]. However, there are a few caveats pertaining to strengthening interventions that are often ignored at the primary care level. In a review of 10 recent randomized trials for strength training, Baker and McAlindon [166] found that only 5 of 10 programs yielded significant improvements in quadriceps strength. If the primary mechanism of weakness is reduced central drive, then it is intuitive that single and multi-joint resistance exercises will have little effect until the cause of reduced central drive is removed. Furthermore, there is no evidence reporting a positive benefit for exercise on structural disease progression. Strength training did not attenuate radiographic disease progression, nor did it alter the knee adduction moment (known to be a surrogate measure for joint loading in medial compartment knee OA) in randomized control trials [167, 168]. One recent controversial study provided evidence that in a malaligned and lax knee joint, increased muscle strength is associated with worse structural progression [169]. Therefore, further research is required to determine the optimal method for rehabilitation of quadriceps muscle weakness and whether quadriceps muscle function has any impact on the disease process.

References

1 Doherty TJ: Invited review: aging and sarcopenia. J Appl Physiol 2003;95:1717–1727.
2 Doherty TJ: The influence of aging and sex on skeletal muscle mass and strength. Curr Opin Clin Nutr Metab Care 2001;4:503–508.
3 Doherty TJ, Vandervoort AA, Brown WF: Effects of ageing on the motor unit: a brief review. Can J Appl Physiol 1993;18:331–358.
4 Rosenberg IH: Summary comments. Am J Clin Nutr 1989;50:1231–1233.
5 Abellan van Kan G, Andre E, Bischoff Ferrari HA, et al: Carla Task Force on Sarcopenia: propositions for clinical trials. J Nutr Health Aging 2009;13:700–707.
6 Porter MM, Vandervoort AA, Lexell J: Aging of human muscle: structure, function and adaptability. Scand J Med Sci Sports 1995;5:129–142.
7 Vandervoort AA: Aging of the human neuromuscular system. Muscle Nerve 2002;25:17–25.
8 Larsson L, Grimby G, Karlsson J: Muscle strength and speed of movement in relation to age and muscle morphology. J Appl Physiol 1979;46:451–456.
9 Murray MP, Duthie EHJ, Gambert SR, Sepic SB, Mollinger LA: Age-related differences in knee muscle strength in normal women. J Gerontol 1985; 40:275–280.
10 Murray MP, Gardner GM, Mollinger LA, Sepic SB: Strength of isometric and isokinetic contractions: knee muscles of men aged 20 to 86. Phys Ther 1980;60:412–419.
11 Young A, Stokes M, Crowe M: The size and strength of the quadriceps muscles of old and young men. Clin Physiol 1985;5:145–154.
12 Young A, Stokes M, Crowe M: Size and strength of the quadriceps muscles of old and young women. Eur J Clin Invest 1984;14:282–287.
13 Lewis SF, Haller RG: Skeletal muscle disorders and associated factors that limit exercise performance. Exerc Sport Sci Rev 1989;17:67–113.
14 Bassey EJ, Harries UJ: Normal values for handgrip strength in 920 men and women aged over 65 years, and longitudinal changes over 4 years in 620 survivors. Clin Sci (Colch) 1993;84:331–337.
15 Cunningham DA, Morrison D, Rice CL, Cooke C: Ageing and isokinetic plantar flexion. Eur J Appl Physiol 1987;56:24–29.
16 Davies CT, Thomas DO, White MJ: Mechanical properties of young and elderly human muscle. Acta Med Scand Suppl 1986;711:219–226.
17 Fisher MB, Birren JE: Age and strength. J Appl Psychol 1947;31:490–497.
18 Kallman DA, Plato CC, Tobin JD: The role of muscle loss in the age-related decline of grip strength: cross-sectional and longitudinal perspectives. J Gerontol 1990;45:M82–M88.
19 McDonagh MJ, White MJ, Davies CT: Different effects of ageing on the mechanical properties of human arm and leg muscles. Gerontology 1984;30: 49–54.
20 Vandervoort AA, McComas AJ: Contractile changes in opposing muscles of the human ankle joint with aging. J Appl Physiol 1986;61:361–367.
21 Campbell MJ, McComas AJ, Petito F: Physiological changes in ageing muscles. J Neurol Neurosurg Psychiatry 1973;36:174–182.
22 Doherty TJ, Brown WF: Age-related changes in the twitch contractile properties of human thenar motor units. J Appl Physiol 1997;82:93–101.
23 Doherty TJ, Vandervoort AA, Taylor AW, Brown WF: Effects of motor unit losses on strength in older men and women. J Appl Physiol 1993;74:868–874.
24 Klein CS, Rice CL, Marsh GD: Normalized force, activation, and coactivation in the arm muscles of young and old men. J Appl Physiol 2001;91:1341–1349.
25 Narici MV, Bordini M, Cerretelli P: Effect of aging on human adductor pollicis muscle function. J Appl Physiol 1991;71:1277–1281.
26 Vandervoort AA, Hayes KC: Plantarflexion muscle function in young and elderly women. Eur J Appl Physiol 1989:389–394.
27 Overend TJ, Cunningham DA, Kramer JF, Lefcoe MS, Paterson DH: Knee extensor and knee flexor strength. J Gerontol 1992;47:M204–M210.
28 Poulin MJ, Vandervoort AA, Paterson DH, Kramer JF, Cunningham DA: Eccentric and concentric torques of knee and elbow extension in young and older men. Can J Sport Sci 1992;17:3–7.
29 Vandervoort AA, Kramer JF, Wharram ER: Eccentric knee strength of elderly females. J Gerontol 1990;45:B125–B128.

30 Fiatarone MA, Marks EC, Ryan ND, Meredith CN, Lipsitz LA, Evans WJ: High-intensity strength training in nonagenarians: effects on skeletal muscle. JAMA 1990;263:3029–3034.
31 Frontera WR, Meredith CN, O'Reilly KP, Knuttgen HG, Evans WJ: Strength conditioning in older men: skeletal muscle hypertrophy and improved function. J Appl Physiol 1988;64:1038–1044.
32 Klitgaard H, Mantoni M, Schiaffino S, et al: Function, morphology and protein expression of ageing skeletal muscle: a cross-sectional study of elderly men with different training backgrounds. Acta Physiol Scand 1990;140:41–54.
33 Overend TJ, Cunningham DA, Paterson DH, Lefcoe MS: Thigh composition in young and elderly men determined by computed tomography. Clin Physiol 1992;12:629–640.
34 Rice CL, Cunningham DA, Paterson DH, Lefcoe MS: Arm and leg composition determined by computed tomography in young and elderly men. Clin Physiol 1989;9:207–220.
35 Lexell J, Taylor CC, Sjostrom M: What is the cause of aging atrophy? J Neurol Sci 1988;84:275–294.
36 Baumgartner RN, Koehler KM, Gallagher D, et al: Epidemiology of sarcopenia among the elderly in New Mexico. Am J Epidemiol 1998;147:755–763.
37 Melton LJr, Khosla S, Crowson CS, O'Connor MK, O'Fallon WM, Riggs BL: Epidemiology of sarcopenia. J Am Geriatr Soc 2000;48:625–630.
38 Iannuzzi-Sucich M, Prestwood KM, Kenny AM: Prevalence of sarcopenia and predictors of skeletal muscle mass in healthy, older men and women. J Gerontol A Biol Sci Med Sci 2002;57:M772–M777.
39 Janssen I, Heymsfield SB, Wang ZM, Ross R: Skeletal muscle mass and distribution in 468 men and women aged 18–88 yr. J Appl Physiol 2000;89:81–88.
40 Janssen I, Heymsfield SB, Ross R: Low relative skeletal muscle mass (sarcopenia) in older persons is associated with functional impairment and physical disability. J Am Geriatr Soc 2002;50:889–896.
41 Delmonico MJ, Harris TB, Visser M, et al: Longitudinal study of muscle strength, quality, and adipose tissue infiltration. Am J Clin Nutr 2009;90: 1579–1585.
42 Hughes VA, Frontera WR, Roubenoff R, Evans WJ, Singh MA: Longitudinal changes in body composition in older men and women: role of body weight change and physical activity. Am J Clin Nutr 2002; 76:473–481.
43 Clark BC, Manini TM: Sarcopenia = / = dynapenia. J Gerontol A Biol Sci Med Sci 2008;63:829–834.
44 Rolland Y, Czerwinski S, Abellan Van Kan G, et al: Sarcopenia: its assessment, etiology, pathogenesis, consequences and future perspectives. J Nutr Health Aging 2008;12:433–450.
45 Doherty T, Simmons Z, O'Connell B, et al: Methods for estimating the numbers of motor units in human muscles. J Clin Neurophysiol 1995;12:565–584.
46 Doherty TJ, Brown WF: Motor unit number estimation: methods and applications; in Brown WF, Bolton CF, Aminoff MJ (eds): Neuromusculoar Function and Disease: Basic, Clinical, and Electrodiagnostic Aspects. Philadelphia, Saunders, 2002, pp 274–290.
47 Grimby G, Saltin B: The ageing muscle. Clin Physiol 1983;3:209–218.
48 Lexell J: Ageing and human muscle: observations from Sweden. Can J Appl Physiol 1993;18:2–18.
49 Lexell J, Downham DY: What determines the muscle cross-sectional area? J Neurol Sci 1992;111:113–114.
50 Roos MR, Rice CL, Vandervoort AA: Age-related changes in motor unit function. Muscle Nerve 1997;20:679–690.
51 Larsson L, Sjodin B, Karlsson J: Histochemical and biochemical changes in human skeletal muscle with age in sedentary males, age 22–65 years. Acta Physiol Scand 1978;103:31–39.
52 Andersen JL, Terzis G, Kryger A: Increase in the degree of coexpression of myosin heavy chain isoforms in skeletal muscle fibers of the very old. Muscle Nerve 1999;22:449–454.
53 Essen-Gustavsson B, Borges O: Histochemical and metabolic characteristics of human skeletal muscle in relation to age. Acta Physiol Scand 1986;126:107–114.
54 Oertel G: Changes in human skeletal muscles due to ageing: histological and histochemical observations on autopsy material. Acta Neuropathol (Berl) 1986; 69:309–313.
55 Brown WF: A method for estimating the number of motor units in thenar muscles and the changes in motor unit count with ageing. J Neurol Neurosurg Psychiatry 1972;35:845–852.
56 Doherty TJ, Brown WF: The estimated numbers and relative sizes of thenar motor units as selected by multiple point stimulation in young and older adults. Muscle Nerve 1993;16:355–366.
57 Stalberg E: Macro EMG, a new recording technique. J Neurol Neurosurg Psychiatry 1980;43:475–482.
58 de Koning P, Wieneke GH, van der Most van Spijk D, van Huffelen AC, Gispen WH, Jennekens FG: Estimation of the number of motor units based on macro-EMG. J Neurol Neurosurg Psychiatry 1988; 51:403–411.
59 Stalberg E, Borges O, Ericsson M, et al: The quadriceps femoris muscle in 20–70-year-old subjects: relationship between knee extension torque, electrophysiological parameters, and muscle fiber characteristics. Muscle Nerve 1989;12:382–389.

60 Brown WF, Strong MJ, Snow R: Methods for estimating numbers of motor units in biceps-brachialis muscles and losses of motor units with aging. Muscle Nerve 1988;11:423–432.
61 Sica RE, McComas AJ, Upton AR, Longmire D: Motor unit estimations in small muscles of the hand. J Neurol Neurosurg Psychiatry 1974;37:55–67.
62 Kawamura Y, O'Brien P, Okazaki H, Dyck PJ: Lumbar motoneurons of man. II. The number and diameter distribution of large- and intermediate-diameter cytons in 'motoneuron columns' of spinal cord of man. J Neuropathol Exp Neurol 1977;36:861–870.
63 Kawamura Y, Okazaki H, O'Brien PC, Dych PJ: Lumbar motoneurons of man. I. Number and diameter histogram of alpha and gamma axons of ventral root. J Neuropathol Exp Neurol 1977;36:853–860.
64 Mittal KR, Logmani FH: Age-related reduction in 8th cervical ventral nerve root myelinated fiber diameters and numbers in man. J Gerontol 1987;42:8–10.
65 Tomlinson BE, Irving D: The numbers of limb motor neurons in the human lumbosacral cord throughout life. J Neurol Sci 1977;34:213–219.
66 Ansved T, Larsson L: Quantitative and qualitative morphological properties of the soleus motor nerve and the L5 ventral root in young and old rats: relation to the number of soleus muscle fibers. J Neurol Sci 1990;96:269–282.
67 Caccia MR, Harris JB, Johnson MA: Morphology and physiology of skeletal muscle in aging rodents. Muscle Nerve 1979;2:202–212.
68 McComas AJ: 1998 ISEK Congress Keynote Lecture: motor units: how many, how large, what kind? International Society of Electrophysiology and Kinesiology. J Electromyogr Kinesiol 1998;8:391–402.
69 McComas AJ: Invited review: motor unit estimation: methods, results, and present status. Muscle Nerve 1991;14:585–597.
70 McNeil CJ, Doherty TJ, Stashuk DW, Rice CL: Motor unit number estimates in the tibialis anterior muscle of young, old, and very old men. Muscle Nerve 2005;31:461–467.
71 Dalton BH, McNeil CJ, Doherty TJ, Rice CL: Age-related reductions in the estimated numbers of motor units are minimal in the human soleus. Muscle Nerve 2008;38:1108–1115.
72 Power G, Dalton BH, Behm DG, Vandervoort AA, Doherty TJ, Rice CL: Motor unit number estimates in masters runners: use it or lose it? Med Sci Sports Exerc 2010;in press.
73 Skelton DA, Greig CA, Davies JM, Young A: Strength, power and related functional ability of healthy people aged 65–89 years. Age Ageing 1994;23:371–377.
74 Metter EJ, Conwit R, Tobin J, Fozard JL: Age-associated loss of power and strength in the upper extremities in women and men. J Gerontol A Biol Sci Med Sci 1997;52:B267–B276.
75 Bean JF, Kiely DK, LaRose S, Leveille SG: Which impairments are most associated with high mobility performance in older adults? Implications for a rehabilitation prescription. Arch Phys Med Rehabil 2008;89:2278–2284.
76 Tipton CM, American College of Sports Medicine: ACSM's Advanced Exercise Physiology. Philadelphia, Lippincott Williams & Wilkins, 2006.
77 Raj IS, Bird SR, Shield AJ: Aging and the force-velocity relationship of muscles. Exp Gerontol 2010;45:81–90.
78 Izquierdo M, Ibanez J, Gorostiaga E, et al: Maximal strength and power characteristics in isometric and dynamic actions of the upper and lower extremities in middle-aged and older men. Acta Physiol Scand 1999;167:57–68.
79 Lanza IR, Towse TF, Caldwell GE, Wigmore DM, Kent-Braun JA: Effects of age on human muscle torque, velocity, and power in two muscle groups. J Appl Physiol 2003;95:2361–2369.
80 Valour D, Ochala J, Ballay Y, Pousson M: The influence of ageing on the force-velocity-power characteristics of human elbow flexor muscles. Exp Gerontol 2003;38:387–395.
81 Toji H, Kaneko M: Effects of aging on force, velocity, and power in the elbow flexors of males. J Physiol Anthropol 2007;26:587–592.
82 McNeil CJ, Vandervoort AA, Rice CL: Peripheral impairments cause a progressive age-related loss of strength and velocity-dependent power in the dorsiflexors. J Appl Physiol 2007;102:1962–1968.
83 Thom JM, Morse CI, Birch KM, Narici MV: Triceps surae muscle power, volume, and quality in older versus younger healthy men. J Gerontol A Biol Sci Med Sci 2005;60:1111–1117.
84 Thom JM, Morse CI, Birch KM, Narici MV: Influence of muscle architecture on the torque and power-velocity characteristics of young and elderly men. Eur J Appl Physiol 2007;100:613–619.
85 Petrella JK, Kim JS, Tuggle SC, Hall SR, Bamman MM: Age differences in knee extension power, contractile velocity, and fatigability. J Appl Physiol 2005;98:211–220.
86 Caserotti P, Aagaard P, Puggaard L: Changes in power and force generation during coupled eccentric-concentric versus concentric muscle contraction with training and aging. Eur J Appl Physiol 2008;103:151–161.

87 De Vito G, Bernardi M, Forte R, Pulejo C, Macaluso A, Figura F: Determinants of maximal instantaneous muscle power in women aged 50–75 years. Eur J Appl Physiol Occup Physiol 1998;78:59–64.
88 Harries UJ, Bassey EJ: Torque-velocity relationships for the knee extensors in women in their 3rd and 7th decades. Eur J Appl Physiol Occup Physiol 1990; 60:187–190.
89 Gajdosik RL, Vander Linden DW, Williams AK: Influence of age on length and passive elastic stiffness characteristics of the calf muscle-tendon unit of women. Phys Ther 1999;79:827–838.
90 Yu F, Hedstrom M, Cristea A, Dalen N, Larsson L: Effects of ageing and gender on contractile properties in human skeletal muscle and single fibres. Acta Physiol (Oxf) 2007;190:229–241.
91 Caserotti P, Aagaard P, Simonsen EB, Puggaard L: Contraction-specific differences in maximal muscle power during stretch-shortening cycle movements in elderly males and females. Eur J Appl Physiol 2001;84:206–212.
92 Lexell J, Taylor CC, Sjostrom M: What is the cause of the ageing atrophy? Total number, size and proportion of different fiber types studied in whole vastus lateralis muscle from 15- to 83-year-old men. J Neurol Sci 1988;84:275–294.
93 Hook P, Sriramoju V, Larsson L: Effects of aging on actin sliding speed on myosin from single skeletal muscle cells of mice, rats, and humans. Am J Physiol Cell Physiol 2001;280:C782–C788.
94 Trappe T: Influence of aging and long-term unloading on the structure and function of human skeletal muscle. Appl Physiol Nutr Metab 2009;34:459–464.
95 D'Antona G, Pellegrino MA, Carlizzi CN, Bottinelli R: Deterioration of contractile properties of muscle fibres in elderly subjects is modulated by the level of physical activity. Eur J Appl Physiol 2007;100:603–611.
96 Krivickas LS, Suh D, Wilkins J, Hughes VA, Roubenoff R, Frontera WR: Age- and gender-related differences in maximum shortening velocity of skeletal muscle fibers. Am J Phys Med Rehabil 2001; 80:447–455; quiz 56–57.
97 Larsson L, Li X, Frontera WR: Effects of aging on shortening velocity and myosin isoform composition in single human skeletal muscle cells. Am J Physiol 1997;272:C638–C649.
98 Ochala J, Frontera WR, Dorer DJ, Van Hoecke J, Krivickas LS: Single skeletal muscle fiber elastic and contractile characteristics in young and older men. J Gerontol A Biol Sci Med Sci 2007;62:375–381.
99 Trappe S, Gallagher P, Harber M, Carrithers J, Fluckey J, Trappe T: Single muscle fibre contractile properties in young and old men and women. J Physiol 2003;552:47–58.
100 Jette AM, Assmann SF, Rooks D, Harris BA, Crawford S: Interrelationships among disablement concepts. J Gerontol A Biol Sci Med Sci 1998;53: M395–M404.
101 Guralnik JM, Simonsick EM, Ferrucci L, et al: A short physical performance battery assessing lower extremity function: association with self-reported disability and prediction of mortality and nursing home admission. J Gerontol 1994;49:M85–M94.
102 Guralnik JM, Ferrucci L, Simonsick EM, Salive ME, Wallace RB: Lower-extremity function in persons over the age of 70 years as a predictor of subsequent disability. N Engl J Med 1995;332:556–561.
103 Wolfson L, Judge J, Whipple R, King M: Strength is a major factor in balance, gait, and the occurrence of falls. J Gerontol A Biol Sci Med Sci 1995;50 Spec No:64–67.
104 Moxley Scarborough D, Krebs DE, Harris BA: Quadriceps muscle strength and dynamic stability in elderly persons. Gait Posture 1999;10:10–20.
105 Corrigan D, Bohannon RW: Relationship between knee extension force and stand-up performance in community-dwelling elderly women. Arch Phys Med Rehabil 2001;82:1666–1672.
106 Salem GJ, Wang MY, Young JT, Marion M, Greendale GA: Knee strength and lower- and higher-intensity functional performance in older adults. Med Sci Sports Exerc 2000;32:1679–1684.
107 Brown M, Sinacore DR, Host HH: The relationship of strength to function in the older adult. J Gerontol A Biol Sci Med Sci 1995:5055–5059.
108 Bassey EJ, Fiatarone MA, O'Neill EF, Kelly M, Evans WJ, Lipsitz LA: Leg extensor power and functional performance in very old men and women. Clin Sci (Lond) 1992;82:321–327.
109 Rantanen T, Avela J: Leg extension power and walking speed in very old people living independently. J Gerontol A Biol Sci Med Sci 1997;52:M225–M231.
110 Bean JF, Kiely DK, Herman S, et al: The relationship between leg power and physical performance in mobility-limited older people. J Am Geriatr Soc 2002;50:461–467.
111 Sayers SP, Guralnik JM, Thombs LA, Fielding RA: Effect of leg muscle contraction velocity on functional performance in older men and women. J Am Geriatr Soc 2005;53:467–471.
112 Foldvari M, Clark M, Laviolette LC, et al: Association of muscle power with functional status in community-dwelling elderly women. J Gerontol A Biol Sci Med Sci 2000;55:M192–M199.
113 Cuoco A, Callahan DM, Sayers S, Frontera WR, Bean J, Fielding RA: Impact of muscle power and force on gait speed in disabled older men and women. J Gerontol A Biol Sci Med Sci 2004;59:1200–1206.

114 Puthoff ML, Janz KF, Nielson D: The relationship between lower extremity strength and power to everyday walking behaviors in older adults with functional limitations. J Geriatr Phys Ther 2008; 31:24–31.
115 Suzuki T, Bean JF, Fielding RA: Muscle power of the ankle flexors predicts functional performance in community-dwelling older women. J Am Geriatr Soc 2001;49:1161–1167.
116 Herman S, Kiely DK, Leveille S, O'Neill E, Cyberey S, Bean JF: Upper and lower limb muscle power relationships in mobility-limited older adults. J Gerontol A Biol Sci Med Sci 2005;60:476–480.
117 Bean JF, Leveille SG, Kiely DK, Bandinelli S, Guralnik JM, Ferrucci L: A comparison of leg power and leg strength within the InCHIANTI study: which influences mobility more? J Gerontol A Biol Sci Med Sci 2003;58:728–733.
118 Mayson DJ, Kiely DK, LaRose SI, Bean JF: Leg strength or velocity of movement: which is more influential on the balance of mobility limited elders? Am J Phys Med Rehabil 2008;87:969–976.
119 Puthoff ML, Nielsen DH: Relationships among impairments in lower-extremity strength and power, functional limitations, and disability in older adults. Phys Ther 2007;87:1334–1347.
120 Ferrucci L, Guralnik JM, Buchner D, et al: Departures from linearity in the relationship between measures of muscular strength and physical performance of the lower extremities: the Women's Health and Aging Study. J Gerontol A Biol Sci Med Sci 1997;52:M275–M285.
121 Earles DR, Judge JO, Gunnarsson OT: Velocity training induces power-specific adaptations in highly functioning older adults. Arch Phys Med Rehabil 2001;82:872–878.
122 Orr R, de Vos NJ, Singh NA, Ross DA, Stavrinos TM, Fiatarone-Singh MA: Power training improves balance in healthy older adults. J Gerontol A Biol Sci Med Sci 2006;61:78–85.
123 Caserotti P, Aagaard P, Larsen JB, Puggaard L: Explosive heavy-resistance training in old and very old adults: changes in rapid muscle force, strength and power. Scand J Med Sci Sports 2008;18:773–782.
124 de Vos NJ, Singh NA, Ross DA, Stavrinos TM, Orr R, Fiatarone Singh MA: Effect of power-training intensity on the contribution of force and velocity to peak power in older adults. J Aging Phys Act 2008; 16:393–407.
125 de Vos NJ, Singh NA, Ross DA, Stavrinos TM, Orr R, Fiatarone Singh MA: Optimal load for increasing muscle power during explosive resistance training in older adults. J Gerontol A Biol Sci Med Sci 2005; 60:638–647.
126 Bean JF, Kiely DK, LaRose S, O'Neill E, Goldstein R, Frontera WR: Increased velocity exercise specific to task training versus the National Institute on Aging's strength training program: changes in limb power and mobility. J Gerontol A Biol Sci Med Sci 2009; 64:983–991.
127 Bottaro M, Machado SN, Nogueira W, Scales R, Veloso J: Effect of high versus low-velocity resistance training on muscular fitness and functional performance in older men. Eur J Appl Physiol 2007; 99:257–264.
128 Katula JA, Rejeski WJ, Marsh AP: Enhancing quality of life in older adults: a comparison of muscular strength and power training. Health Qual Life Outcomes 2008;6:45.
129 Marsh AP, Miller ME, Rejeski WJ, Hutton SL, Kritchevsky SB: Lower extremity muscle function after strength or power training in older adults. J Aging Phys Act 2009;17:416–443.
130 Fielding RA, LeBrasseur NK, Cuoco A, Bean J, Mizer K, Fiatarone Singh MA: High-velocity resistance training increases skeletal muscle peak power in older women. J Am Geriatr Soc 2002;50:655–662.
131 Henwood TR, Riek S, Taaffe DR: Strength versus muscle power-specific resistance training in community-dwelling older adults. J Gerontol A Biol Sci Med Sci 2008;63:83–91.
132 Reid KF, Callahan DM, Carabello RJ, Phillips EM, Frontera WR, Fielding RA: Lower extremity power training in elderly subjects with mobility limitations: a randomized controlled trial. Aging Clin Exp Res 2008;20:337–343.
133 Guccione AA, Felson DT, Anderson JJ, et al: The effects of specific medical conditions on the functional limitations of elders in the Framingham Study. Am J Public Health 1994;84:351–358.
134 Baker KR, Xu L, Zhang Y, et al: Quadriceps weakness and its relationship to tibiofemoral and patellofemoral knee osteoarthritis in Chinese: the Beijing osteoarthritis study. Arthritis Rheum 2004;50:1815–1821.
135 Felson DT: Clinical practice. Osteoarthritis of the knee. N Engl J Med 2006;354:841–848.
136 McAlindon TE, Cooper C, Kirwan JR, Dieppe PA: Determinants of disability in osteoarthritis of the knee. Ann Rheum Dis 1993;52:258–262.
137 O'Reilly SC, Jones A, Muir KR, Doherty M: Quadriceps weakness in knee osteoarthritis: the effect on pain and disability. Ann Rheum Dis 1998;57:588–594.
138 Hassan BS, Mockett S, Doherty M: Static postural sway, proprioception, and maximal voluntary quadriceps contraction in patients with knee osteoarthritis and normal control subjects. Ann Rheum Dis 2001;60:612–618.

139 Hurley MV, Newham DJ: The influence of arthrogenous muscle inhibition on quadriceps rehabilitation of patients with early, unilateral osteoarthritic knees. Br J Rheumatol 1993;32:127–131.

140 Hurley MV, Scott DL, Rees J, Newham DJ: Sensorimotor changes and functional performance in patients with knee osteoarthritis. Ann Rheum Dis 1997;56:641–648.

141 Lewek MD, Rudolph KS, Snyder-Mackler L: Quadriceps femoris muscle weakness and activation failure in patients with symptomatic knee osteoarthritis. J Orthop Res 2004;22:110–115.

142 Ling SM, Conwit RA, Talbot L, et al: Electromyographic patterns suggest changes in motor unit physiology associated with early osteoarthritis of the knee. Osteoarthritis Cartilage 2007.

143 Hall KD, Hayes KW, Falconer J: Differential strength decline in patients with osteoarthritis of the knee: revision of a hypothesis. Arthritis Care Res 1993;6: 89–96.

144 Hortobagyi T, Garry J, Holbert D, Devita P: Aberrations in the control of quadriceps muscle force in patients with knee osteoarthritis. Arthritis Rheum 2004;51:562–569.

145 Slemenda C, Brandt KD, Heilman DK, et al: Quadriceps weakness and osteoarthritis of the knee. Ann Intern Med 1997;127:97–104.

146 Fisher NM, Pendergast DR: Reduced muscle function in patients with osteoarthritis. Scand J Rehab Med 1997;29:213–221.

147 Slemenda C, Heilman DK, Brandt KD, et al: Reduced quadriceps strength relative to body weight: a risk factor for knee osteoarthritis in women? Arthritis Rheum 1998;41:1951–1959.

148 Becker R, Berth A, Nehring M, Awiszus F: Neuromuscular quadriceps dysfunction prior to osteoarthritis of the knee. J Orthop Res 2004;22:768–773.

149 Brandt KD, Radin EL, Dieppe PA, van de Putte L: Yet more evidence that osteoarthritis is not a cartilage disease. Ann Rheum Dis 2006;65:1261–1264.

150 Jefferson RJ, Collins JJ, Whittle MW, Radin EL, O'Connor JJ: The role of the quadriceps in controlling impulsive forces around heel strike. Proc Inst Mech Engn 1990;204:21–28.

151 Radin EL, Yang KH, Riegger C, Kish VL, O'Connor JJ: Relationship between lower limb dynamics and knee joint pain. J Orthop Res 1991;9:398–405.

152 Rudolph KS, Schmitt LC, Lewek MD: Age-related changes in strength, joint laxity, and walking patterns: are they related to knee osteoarthritis? Phys Ther 2007;87:1422–1432.

153 Schmitt LC, Rudolph KS: Influences on knee movement strategies during walking in persons with medial knee osteoarthritis. Arthritis Rheum 2007; 57:1018–1026.

154 Longino D, Frank C, Herzog W: Acute botulinum toxin-induced muscle weakness in the anterior cruciate ligament-deficient rabbit. J Orthop Res 2005; 23:1404–1410.

155 Rehan Youssef A, Longino D, Seerattan R, Leonard T, Herzog W: Muscle weakness causes joint degeneration in rabbits. Osteoarthritis Cartilage 2009;17: 1228–1235.

156 Odding E, Valkenburg HA, Algra D, Vandenouweland FA, Grobbee DE, Hofman A: Associations of radiological osteoarthritis of the hip and knee with locomotor disability in the Rotterdam Study. Ann Rheum Dis 1998;57:203–208.

157 Creamer P, Lethbridge-Cejku M, Hochberg MC: Factors associated with functional impairment in symptomatic knee osteoarthritis. Rheumatology (Oxford, England) 2000;39:490–496.

158 Ling SM, Xue QL, Simonsick EM, et al: Transitions to mobility difficulty associated with lower extremity osteoarthritis in high functioning older women: longitudinal data from the Women's Health and Aging Study II. Arthritis Rheum 2006;55:256–263.

159 Ikeda S, Tsumura H, Torisu T: Age-related quadriceps-dominant muscle atrophy and incident radiographic knee osteoarthritis. J Orthop Sci 2005;10: 121–126.

160 Petterson SC, Barrance P, Buchanan T, Binder-Macleod S, Snyder-Mackler L: Mechanisms underlying quadriceps weakness in knee osteoarthritis. Med Sci Sports Exerc 2008;40:422–427.

161 Berth A, Urbach D, Awiszus F: Improvement of voluntary quadriceps muscle activation after total knee arthroplasty. Arch Phys Med Rehab 2002;83:1432–1436.

162 Stevens JE, Mizner RL, Snyder-Mackler L: Quadriceps strength and volitional activation before and after total knee arthroplasty for osteoarthritis. J Orthop Res 2003;21:775–779.

163 Heiden TL, Lloyd DG, Ackland TR: Knee extension and flexion weakness in people with knee osteoarthritis: is antagonist cocontraction a factor? J Orthop Sports Phys Ther 2009;39:807–815.

164 Lewek MD, Ramsey DK, Snyder-Mackler L, Rudolph KS: Knee stabilization in patients with medial compartment knee osteoarthritis. Arthr Rheum 2005;52:2845–2853.

165 Recommendations for the Medical Management of Osteoarthritis of the Hip and Knee: 2000 update. American College of Rheumatology Subcommittee on Osteoarthritis Guidelines. Arthr Rheum 2000;43: 1905–1915.

166 Baker K, McAlindon T: Exercise for knee osteoarthritis. Curr Opin Rheumatol 2000;12:456–463.

167 Mikesky AE, Mazzuca SA, Brandt KD, Perkins SM, Damush T, Lane KA: Effects of strength training on the incidence and progression of knee osteoarthritis. Arthritis Rheum 2006;55:690–699.
168 King LK, Birmingham TB, Kean CO, Jones IC, Bryant DM, Giffin JR: Resistance training for medial compartment knee osteoarthritis and malalignment. Med Sci Sports Exerc 2008;40:1376–1384.
169 Sharma L, Dunlop DD, Cahue S, Song J, Hayes KW: Quadriceps strength and osteoarthritis progression in malaligned and lax knees. Ann Intern Med 2003;138:613–619.

Dr. T.J. Doherty
RM C7–131, University Hospital, LHSC
339 Windermere Rd.
London, ON N6A 5A5 (Canada)
Tel. +1 519 663 3337, E-Mail tim.doherty@lhsc.on.ca

Mobbs CV, Hof PR (eds): Body Composition and Aging.
Interdiscipl Top Gerontol. Basel, Karger, 2010, vol 37, pp 115–141

mTOR Signaling as a Target of Amino Acid Treatment of the Age-Related Sarcopenia

Giuseppe D'Antona[a] · Enzo Nisoli[b]

[a]Department of Physiology, Human Physiology Unit and Interuniversity Institute of Myology, University of Pavia, Pavia, and [b]Department of Pharmacology, Chemotherapy and Medical Toxicology, University of Milan, Milan, Italy

Abstract

Sarcopenia is an age-related structural and functional impairment of skeletal muscle leading to loss of strength, contractile capacity and endurance. Among factors implicated in sarcopenia, deregulation of muscle protein synthesis (MPS) has frequently been reported. Thus, the attempts aiming at identifying possible countermeasures to sarcopenia require consideration of a complex coordinated interaction of factors contributing to the balance between protein synthesis and breakdown and the identification of several regulators on their function. We will focus here on the signaling pathways controlling protein synthesis in skeletal muscle, specifically on one of the downstream effectors of the kinase Akt/PKB, the mammalian target of rapamycin (mTOR) kinase which is now recognized as a key regulator of cell growth and a pivotal sensor of nutritional status over the lifespan. Dysfunction of mTOR signaling in the elderly and its potential role as a target of amino acids in the treatment of age-related sarcopenia will be discussed.

Skeletal muscle represents up to 45% of body weight of adult humans [1] and for its mass to be stable strictly depends on equal rates of protein synthesis and breakdown. Under physiological conditions, continuous protein turnover is a basic cellular process that serves to maintain the conservation of cytoplasm content and cell size and to provide amino acids for oxidative and biosynthetic reactions. As a direct consequence, the balance between new protein formation (synthesis) and catabolism (breakdown) determines the overall protein content within each muscle fiber [2]. The synthetic rates of the two major myofibrillar proteins (i.e. myosin and actin) and the noncontractile proteins, including sarcoplasmic proteins and others, have been extensively investigated in several species and in numerous conditions [3–6]. In healthy adult humans, rates of myofibrillar protein synthesis have been calculated (around 0.8 g day^{-1} kg^{-1}) [4] and the mean daily turnover rate has been expressed as a fraction of the total body protein turnover, which ranges in different studies from less than 1% to more than 2% [4–6], without differences between men and women [5, 7, 8]. This fraction, which appears to be much higher during development (more than 3% in newborns and 5% in premature newborns) [9–11], is an index of muscle growth [10] and correlates with the age-related increase in fiber size. In particular, the protein synthetic rates and muscle growth are positively related during the period of life in which muscle weight is proportional to body weight [10], while this positive relationship appears to be lost in presence of quantitative reduction of skeletal muscle mass (atrophy). This loss physiologically occurs in healthy aging, and is known as sarcopenia, which is defined in humans by appendicular skeletal muscle mass per height^2 (<7.26 kg m^{-2} for males and <5.45 kg m^{-2} for females) [12, 13]. In fact, in the presence of sarcopenia, which is responsible for unequal loss of muscle mass and force generation [14, 15], increased prevalence of falls and fractures, greater morbidity and loss of autonomy in the elderly is associated with lower myofibrillar protein turnover and reduced protein synthesis, increased protein degradation, or a combination of both [16]. Furthermore, a clear unbalance between protein synthesis and breakdown appears as the *leitmotif* of aging and several other conditions leading to severe muscle wasting, including uncontrolled metabolic disorders (diabetes) [7, 17], denervation [7, 18], atrophic myopathies [19], dystrophies [9, 20], sepsis [21–24] and cancer [25–27]. Considering the shared endpoint of different atrophic conditions, common mechanisms able to affect synthesis and/or degradation might exist and several coexisting regulatory factors such as hormones (growth hormone, insulin-like growth factor-1 or IGF-1, insulin, testosterone, glucocorticoids, thyroxine), nutritional status (amino acids availability), and the basal level of physical activity [15, 28] may have differential impact on metabolic balance. In any case several lines of evidence suggest that muscle atrophy, and thus sarcopenia, is not the simple converse of hypertrophy [29], and distinct pathways and unique set of genes are differentially activated [30, 31]. The onset of skeletal muscle atrophy is strictly related to the activation of the ATP-dependent ubiquitin-proteasome pathway [32] and follows the increased levels of mRNAs encoding for essential components of the ubiquitin-proteasome machinery [31, 32]. The inhibition of this pathway decreases protein breakdown in skeletal muscles, thus preventing proteolysis [33]. Furthermore, in experiments to identify specific markers of muscle wasting, two genes, atrogin-1/MAFbx and MuRF1, appeared significantly upregulated before the onset of mass loss in different models of muscle atrophy. These genes encode for muscle-specific ubiquitin ligases responsible for substrate specificity during protein degradation through the ubiquitin-proteasome pathway [34–36]. The demonstration that ablation of either gene results in protection against denervation-induced atrophy, highlights the relevance of these two ligases in the onset of skeletal muscle atrophy [34].

On the other hand, in the presence of physiological conditions promoting muscle growth, a major role in controlling the cellular hypertrophic response has to be attributed to the activation of PI3K-Akt signaling pathway due to an increased expression of IGF-1. IGF-1, the best characterized muscle growth factor, is synthesized in the liver and, locally, in the skeletal muscle [37] where its overexpression in transgenic mice results in fiber hypertrophy [38, 39]. Muscle expression of IGF-1 has been found to increase following compensatory overload [40, 41]. The bulk of evidence supports the crucial role of protein kinase B (Akt/PKB) pathway in the regulation of fiber size. The acute activation of Akt/PKB kinase promotes highly significant fiber hypertrophy in vivo [42] whereas decreased ability to activate Akt/PKB is coupled with atrophy [43].

The recent identification of a family of transcription factors, the Forkhead box O (FoxOs), negatively targeted by Akt/PKB has opened new vistas on the coordinated response of different signaling pathways controlling atrophy and hypertrophy [44, 45]. In particular, Akt/PKB promotes FoxO phosphorylation followed by its translocation from the nucleus to the cytoplasm. The inverse translocation of FoxO, from the cytoplasm to the nucleus, is necessary for upregulation of atrogin-1/MAFbx and MuRF1 [44] and this mediates the development of atrophy. Activation of the PI3K/Akt pathway by IGF-1 prevents muscle loss through suppression of atrogin-1/MAFbx and MuRF1 expression following inactivation of FoxO [44]. In contrast, in presence of atrophy, reducing FoxO expression by RNAi prevents atrogin-1/MAFbx upregulation [44]. These findings highlight the existence of synergic responses of cellular mediators of hypertrophy and atrophy to a stimulus, such as the local increase/decrease of IGF-1 in skeletal muscle, and suggest future intriguing directions of research. In particular, attempts to identify possible countermeasures to atrophy/sarcopenia of aging [15, 46–48] require consideration of a complex coordination of factors contributing to the control of cell size and the impact of several regulators on their function. We will focus here on the signaling pathways controlling protein synthesis in skeletal muscle, specifically on one of the downstream effectors of the Akt/PKB, the mammalian target of rapamycin (mTOR) kinase which is now recognized as a key regulator of cell growth and a pivotal sensor of nutritional status over the lifespan. The role of mTOR as a potential target of amino acids and its possible involvement in amino acid-induced improvement of age-related sarcopenia will be discussed.

Intracellular Mechanisms of Protein Synthesis Regulation: mTOR as a Downstream Target of PKB/Akt

Insight into the intracellular mechanisms regulating fibers size mostly derives from initial findings in *Drosophila*, in which the overexpression of Akt/PKB was associated with hypertrophy [49–52]. Akt (PKB/Akt) is a serine/threonine kinase, which in mammals consists of the three highly homologous isoforms Akt1 (PKBa), Akt2 (PKBb), and Akt3 (PKBg) [53]. In skeletal muscle Akt1 and Akt2 are more highly expressed than Akt3. Each PKB/Akt isoform has an N-terminal pleckstrin homology (PH) domain [54] which mediates binding of PKB/Akt to 3-phosphoinositides and is followed by the kinase catalytic domain. In this region there is a threonine residue

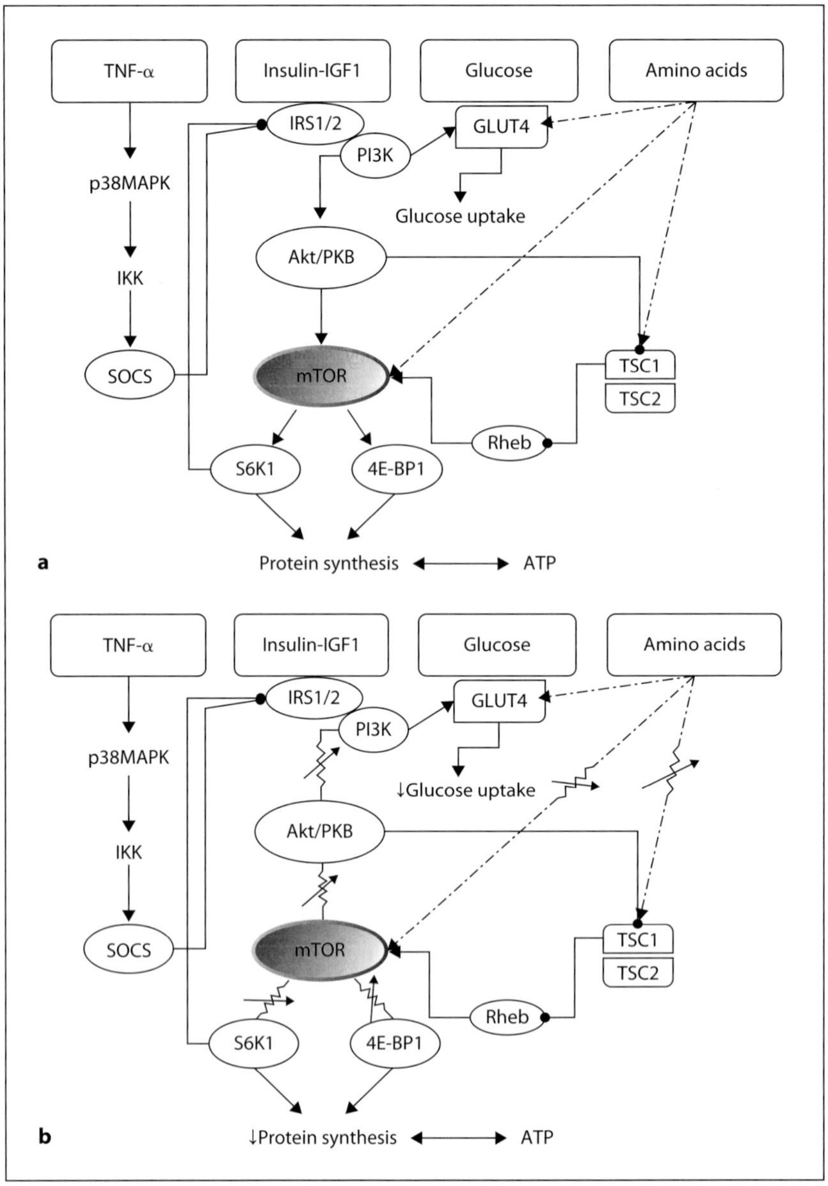

Fig. 1. Schematic depiction of intracellular regulation of protein synthesis via mTOR signaling and other routes in the young (**a**) and the elderly (**b**). **a** Insulin and IGF-1 activate a cascade of events leading to phosphorylation PI3 and PKB/Akt followed by activation of mTOR which increases its activity on S6K1 and 4E-BP1 that promote mRNA translation initiation and protein synthesis. S6K1 inhibits IRS1 and contributes to signal extinction and arising of insulin resistance. Insulin resistance may also arise from activation of SOCS by TNF-α mediated by IKK. Akt/PKB phosphorylates TSC2 on TSC1/2 complex acting on Rheb which accumulates in its GTP bound form on mTORC1 thus reliving inhibition of mTOR. Dashed lines indicate signaling connections that remain to be elucidated. In particular, the mechanisms by which amino acids directly or indirectly regulate mTOR, and, possibly, GLUT4 are unknown. In the elderly (**b**) multiple sites of variable resistance impair the efficiency of signal pathways to various stimuli. In particular insulin resistance of aging is mediated by increased IRS1 phosphorylation and is followed by reduced glucose cell uptake and reduced insulin-mediated

(Thr308 in PKBa/Akt1) whose phosphorylation is necessary for PKB/Akt activation [55, 56]. Next to the kinase domain is a C-terminal tail containing a regulatory phosphorylation site (Ser473 in PKBa/Akt1) whose phosphorylation is essential for maximal PKB/Akt activation [57].

The importance of PKB/Akt in cell growth was initially suspected when it was shown to be a downstream target of phosphatidylinositol 3-kinase (PI3K) [58, 59]. The crucial role of Akt/PKB was subsequently confirmed by the observation of fiber hypertrophy and the attenuation of denervation-induced atrophy following its overexpression in skeletal muscle in vivo [60, 61]. Reduced PKB/Akt activation was also observed in unloaded muscles [60].

Upstream PKB/Akt activation depends on various stimuli, including mechanical load and hormones. Several lines of evidence suggest that contraction, functional overload and passive stretch of fast muscles [60, 62–64] lead to increased PKB/Akt activity, although it is not currently known how the active or passive mechanical load is transduced to signal activation. The hormonal stimulation represents another crucial trigger of the PKB/Akt phosphorylation under physiological conditions. In particular, insulin activates Akt2 while IGF-1 activates Akt1 [65]. Interaction between insulin/IGF-1 and insulin receptor (IR) is followed by modulation of a series of intracellular signaling cascades, including PI3K [53, 65, 66] and, subsequently, PKB/Akt. In fact, PI3K activation determines the synthesis of phosphatidylinositol 3,4,5-triphosphates which recruit PKB/Akt to the plasma membrane. This link is followed by PKB/Akt phosphorylation mediated by two distinct kinases (PDK1 and mTOR-rictor, rapamycin insensitive companion of mTOR complex) [67].

The two major downstream targets of PKB/Akt are mTOR (fig. 1), which controls protein synthesis and plays a key role in sensing the amino acid concentration in the cytoplasm, and FoxO, which controls protein degradation. mTOR expression and function is essential for life. In fact, mice lacking mTOR die in utero shortly after implantation [68]. mTOR was first discovered as a target of rapamycin, a bacterial natural product used as potent immunosuppressant. Next a single gene encoding for mTOR (*TOR)* has been identified in every eukaryote genome examined so far unlike yeast, which, in some cases, possess two *TOR* genes [69, 70]. mTOR belongs to a group of kinases known as the phosphatidylinositol kinase-related kinases (PIKK). PIKK

activation of mTOR. Furthermore, reduced response of mTOR signaling to amino acids leads to further loss of translation initiation and protein synthesis. The existence of insulin-independent mechanisms of mTOR activation may promote stimulation of translation through increased amino acids availability even in presence of age-related insulin resistance. IGF1 = Insulin-like growth factor; IRS1 = insulin receptor substrate 1; PI3 = phosphatidylinositol 3-kinase; PKB/Akt = protein kinase B; mTOR = mammalian target of rapamycin; 4E-BP1 = eIF4E-binding protein; S6K1 = s6 kinase; TSC1/2 tuberous sclerosis complex; Rheb = Ras homolog enriched in the brain; TNF-α = tumor necrosis factor-α; MAPK = mitogen-activated protein kinase; IKK = IKB kinase; SOCS = suppressor of cytokine signaling; GLUT-4 = glucose transporter 4.

family members contain a carboxy-terminal serine/threonine protein kinase domain and interact with protein partners, including scaffolds and regulators. Rapamycin inhibits mTOR through the association with FKBP12 intracellular receptor [71]. The amino-terminal to the kinase domain in mTOR is the FKBP12-rapamycin-binding domain and this interaction inhibits several, but not all, functions of mTOR. In cells, mTOR forms two distinct complexes, mTORC1 and mTORC2, depending on the binding partners. When bound to raptor (regulatory-associated protein of mTOR) mTOR forms mTORC1, which mediates the effects sensitive to rapamycin [72, 73].

mTORC1 regulates cell growth, protein synthesis, and autophagy through phosphorylation of downstream effectors including translation regulators, such as S6K1 and 4E-BP1 [74, 75]. As a consequence, mTOR inhibition prevents the mechanical overload-induced changes in size of both slow and fast skeletal muscle fibres [60].

mTOR and mRNA Translation. Protein synthesis proceeds through the complex process of mRNA translation which includes three consecutive steps: (1) initiation, in which the initiator methionyl-tRNA and mRNA bind to 40S ribosomal subunit; (2) elongation, by which tRNA-bound amino acids are incorporated into growing polypeptide chains according to the mRNA template, and (3) termination, where the completed protein is released from the ribosome [76].

The first two steps of mRNA translation are thought to be highly regulated at two different levels: the binding of methionyl-tRNA to 40S ribosomal subunit to form 43S preinitiation complex, and recognition, unwinding, and binding of mRNA to the 43S, catalyzed by a multi-subunit complex of eukaryotic factors (eIFs), referred to as eIF4F [76–78].

mTOR-mediated regulation of protein synthesis is based on the interaction between raptor, S6K1, and 4E-BP1 at signaling motifs (TOS), followed by phosphorylation of both kinases [79]. Activation of elF4E-binding protein-1 (4E-BP1) releases the inhibition on the eukaryotic factors complex eIF4F, which is responsible for the interaction with 40s ribosomal subunit and translation initiation [80]. In fact when 4E-BP1 is in its hypophosphorylated state, it blocks the ability of eIF4E to bind to eIF4G and forms an inactive 4E-BP1-eIF4E complex. This interaction precludes mRNA to bind to the ribosome [81, 82]. Interestingly, an increased amount of 4E-BP1 mRNA and 4E-BP1 protein bound to eIF4E to form the inactive 4E-BP1/eIF4E complex [60] has been observed in unloaded skeletal muscles [36].

S6K1 is a kinase which requires phosphorylation at two sites and its activation is necessary for muscle fibers to achieve normal size, since S6K1 knockout cells are smaller than control cells [83]. Following phosphorylation at Thr389 by mTORC1, and phosphorylation at of the T-loop by phosphoinositide-dependent kinase-1 (PDK1), S6K1 regulates the activity of eukaryotic elongation factor 2 kinase (eEF2k) [84]. eEF2k inactivates eEF2 in a rapamycin/insulin-dependent manner [85] thus promoting an overall reduction of the ribosomal capacity [84]. It is possible that S6K1 should be primed first by mTORC1 and then activated by PDK1 [86, 87]. However, one can not rule out that S6K1 can be activated independently of mTOR

[88]. Interestingly, muscle unloading appears to be associated with decreased mTOR and S6K1 activation, consistent with reduced translation rate and the appearance of disuse atrophy [60, 89]. Furthermore, a feedback loop involving phosphorylation of mTOR at Thr2446 and Ser2448 by S6K1 has been reported but its role in mTORC1 activation/inactivation is not currently known [90, 91].

Cross-Talk between Insulin and mTOR Pathways. There has been recent progress in the elucidation of the signaling cascade downstream IR stimulation, and the reciprocal cross-relationship between insulin-IRS1-PI3K-PDK1-Akt pathway and mTOR-S6K1 pathway. This suggests an important role of mTOR in controlling glucose homeostasis and protein synthesis through insulin dependent and insulin independent mechanisms [80]. Glucose uptake by insulin is mediated by PI3K-dependent and -independent pathways [92, 93] (fig. 1). Binding of insulin to its receptor leads to the autophosphorylation of transmembrane IR subunits and the tyrosine phosphorylation of the IRS, which interact with the p85 subunit of PI3K [90]. PI3K activation is negatively modulated by PIP3 dephosphorylation via 3′-phosphatases (PTEN and SHIP2) [94, 95]. After phosphorylation of Tyr residues by IR stimulation, IRS1 plays the role of a docking site for many effector proteins bearing an SH2 domain leading to the transmission of the insulin signal [96]. Modulation of PI3K activity by IRS1 is responsible for Akt/PKB activation [97]. In fact, PIP3 mediates translocation of Akt/PKB to the plasma membrane, via its PH domain, and formation of the membrane-bound form of Akt/PKB results in increased glucose transport [98]. Furthermore, the mTOR-rictor complex activates Akt through phosphorylation on Ser473 [99].

Several proteins have the potential to regulate mTOR and the tumor suppressor complex plays a major role [76]. In particular, TSC1 and TSC2 gene products, hamartin and tuberlin, interact to form a stable tumor suppressor complex and mutations in either TSC1 or TSC2 gene cause tuberous sclerosis, an autosomal-dominant disorder characterized by a broad spectrum of symptoms including seizures, mental retardation and renal dysfunction [100].

TSC1/TSC2 complex inhibits mTOR-mediated signaling to 4E-BP1 and S6K1 (fig. 1). TSC2 has a GTPase-activating protein domain acting on the Ras family GTPase called RAS homolog enriched in brain (Rheb) [101–103]. Rheb activates mTOR through phosphorylation on Ser2448- and TSC2-dependent stimulation of GTP hydrolysis of Rheb inhibits mTOR activation on Ser 2448. Thus, TSC1/TSC2 complex is an mTOR suppressor which attenuates mTOR dependent activation of 4E-BP1 and S6K1. Also TSC2 activity is regulated through phosphorylation at different Ser, Tyr and Thr residues [104–107]. In particular, phosphorylation at Thr1462 by Akt/PKB causes dissociation of the TSC1/TSC2 complex, thus relieving inhibition of mTOR through limitation of Rheb-GTPase activity [104] (fig. 1).

Recent reports show that the mechanism for mTOR activation by insulin involves the TSC1/2-Rheb complex [108, 109]. The loss of function of this complex implicates a constitutively active signaling through the mTOR pathway. On the other hand, the inappropriate activation of the TSC-Rheb-mTOR pathway induces IRS1/2 depletion, insulin resistance and survival deficiencies [110].

Although an extensive cross-talk exists between insulin and mTOR signaling, numerous studies suggest that a defect in mTOR-mediated translation does not account for the appearance of insulin resistance in diabetes. In fact, in diabetic animal models a rapamycin-sensitive pathway is a major mediator of the stimulatory effect of insulin on muscle protein synthesis (MPS) [111], and insulin administration is followed by recovery of mTOR-mediated translation initiation through 4E-PB1 and S6K1 activation [112, 113].

The current understanding of the regulation of insulin signaling mediated by mTOR focuses on the function of IRS1, whose phosphorylation on different Ser residues enhances or blocks insulin signaling, while phosphorylation on different Tyr residues activates the response to insulin [114].

The main mechanism of insulin desensitization involves phosphorylation of the Ser/Thr residues on IRS1 and its reduced Tyr phosphorylation [115]. In fact, in the rat, phosphorylation of the Tyr residues 608/628 after insulin stimulation is necessary for the interaction with PI3K and for the subsequent propagation of the signal. On the contrary, phosphorylation of the Ser/Thr residue 307 (e.g. by IKK) leads to a reduced insulin signaling whereas phosphorylation on the Ser/Thr residue 302 enhances or reduces insulin signaling [114, 116, 117].

Also, phosphorylation at Ser 636/639 is a known mechanism that negatively modulates insulin signaling [118, 119].

A large number of protein kinases cause Ser phosphorylation of IRS, including S6K1, which are responsible for development of insulin resistance [120]. The negative feedback mechanism mediated by S6K1 activation on IRS1 (fig. 1) appears to be relevant in physiological conditions and in the development of metabolic disorders such as diabetes and obesity. Interestingly, physiological hyperinsulinemia leads to S6K1 activation [121] and the homeostatic negative-feedback loop may contribute to the signal extinction [122] (fig. 1). IRS1 phosphorylation through activation of S6K1 can also be directly triggered by acute infusion of amino acids and this effect is reversed by rapamycin. In fact, in humans, amino acid infusion increases IRS1 phosphorylation at Ser312 and Ser636/639 and decreases insulin-induced phosphoinositide 3-kinase activity [115]. Considering that amino acid infusion stimulates translation via phosphorylation and activation of S6K1 in a rapamycin-dependent manner, and inhibits insulin secretion under conditions of euglycemia [123], the negative feedback loop mechanism may prevent redundancy of stimuli on the signal transduction pathways involved.

However, the role of S6K1 in mediating IRS1 phosphorylation in settings of chronic nutrient supplementation is currently unknown. Prolonged administration of a branched-chain enriched amino acid mixture (BCAAm) resulted in increased insulin sensitivity in aged and diabetic humans (type 2 diabetes) [124, 125], an observation potentially related to an increased expression and autophosphorylation of IR as observed in experimental and clinical type 2 diabetes [126, 127]. However, no change in the level of IRS1 Ser636 phosphorylation was observed after prolonged BCAAm administration to elderly rats [D'Antona and Nisoli, unpubl. results].

IRS1 function is negatively regulated by molecular mechanisms other than desensitization, including degradation through ubiquitine proteasome pathway [128–131] and inhibition by SOCS (suppressor of cytokine signaling) proteins [132]. After prolonged insulin stimulation, Ser phosphorylation of IRS1 causes its dissociation from the IR and association with MDM2, an E3 ubiquitin ligase that can ubiquitinate IRS1, thus inducing its degradation by the 26S proteasome [128–131]. Interestingly, recent evidence suggests that prolonged supplementation of BCAAm to elderly rats restores insulin-dependent degradation of IRS1 to the adult pattern in in vitro assays [D'Antona and Nisoli, unpubl. results]. The mechanism of IRS1 inhibition mediated by SOCS proteins has recently drawn much attention (fig. 1). The SOCS family comprises eight members described as negative regulators of cytokine signaling [133–135]. SOCS act through several mechanisms including inactivation of kinases, facilitation of proteosomal degradation of signaling mediators and the occupation of docking sites. It has been found that SOCS genes are induced by insulin treatment and their overexpression is followed by insulin resistance mainly through the impairment of IRS phosphorylation. In fact, SOCS1 and SOCS3 cause insulin resistance through the inhibition of IRS phosphorylation [132].

Much More than Bricks: Branched-Chain Amino Acids as Possible Regulators of Protein Synthesis through the mTOR Pathway

In healthy humans approximately 250–350 g of muscle proteins are degraded every day [136]. Although amino acids deriving from continual protein turnover are mainly reused to synthesize new cytoplasmic and mitochondrial proteins or to produce energetic intermediates such as glucose in the liver, a large amount of amino acids is released into the blood and constitutes the blood pool. Studies in animals models and humans showed that free essential amino acids (EAA) [137, 138], unlike non EAA [137], are not only building bricks of proteins or source of energetic intermediates but also fundamental regulators of MPS.

In the young [139] and in the elderly [140] hyperaminoacidemia stimulates amino acid transport and net MPS, unlike carbohydrate administration, which is followed by increased plasma insulin concentration [141]. The effect on protein synthesis arises independent of changes in anabolic hormone concentration [142, 143], although physiological insulin availability is required for the effects of amino acids on translation [113, 144].

Recent studies demonstrate that among EAA, BCAAs (leucine, isoleucine and valine) and, in particular, free leucine play a very important role in protein metabolism [145, 146]. In fact, beside its role as an insulin secretagogue [147], leucine appears as a nutrient signal that regulates protein synthesis through the stimulation of insulin-independent and rapamycin-sensitive pathways [148, 149]. Interestingly, when compared to amino acid mixtures including essential and nonessential amino acids, BCAA or leucine alone appear able to reproduce the same effects of the mixtures on MPS, thus suggesting that the anabolic effect strictly depends on a specific signal initiated by leucine/BCAA [150, 151]. Further observations have strengthened the idea that the effects of the BCAA mixtures on MPS may be attributable only to leucine, at least in vitro. In particular, Buse and Reid [152] and Li and Jefferson [153] first showed that leucine stimulates MPS similarly to a mixture of all BCAA. Although some data indicate that the stimulatory action of leucine is independent of other BCAA in vitro, contradictory results have been obtained in vivo [152, 154], and discrepancies may be related to the role of specific combinations of BCAA in promoting MPS.

The rate of MPS depends on ribosomal efficiency. Early experiments demonstrated that in the postprandial period a significant increase in translational efficiency arises despite no change in ribosomal content [155–157]. It is now clear from the available data that at least part of the postprandial translational activation is to be attributed to BCAA through activation of the mTOR signaling.

The bulk of evidence suggests that leucine administration markedly increases MPS concomitant with the hyperactivation of elF4G, S6K1 and 4E-BP1. Interestingly, hyperphosphorylation of S6K1 and 4E-BP1 may also follow isoleucine administration although with no concomitant increase of MPS [141].

The role of mTOR in nutrient-mediated hyperactivation of 4E-BP1 and S6K1 has been investigated through the analysis of the signaling pathway to rapamycin. Although rapamycin treatment is responsible for reduced protein synthesis independently of leucine administration [158], it completely prevents 4E-BP1 and S6K1 activation mediated by leucine. Furthermore, rapamycin pretreatment (2 h before leucine administration) reduces leucine-dependent activation of mTOR, elF4E and S6K1 and releases elF4E from the 4E-BP1-elF4E complex in the rat [141]. However, the inefficiency of rapamycin to completely block the mTOR signaling activated by leucine suggests that this pathway activation is not sufficient to fully explain the activation of BCAA-dependent anabolism in postabsorptive conditions [76] and alternative pathways might exist. Furthermore, the mTOR pathway also plays an important role in controlling amino acid transportation across the plasma membrane which can represent an important limiting factor of the anabolic response to nutrients [159].

Although mTOR signaling plays the role of integrator of the anabolic response to BCAA, mTOR itself may not be the direct target of amino acids. In fact, recent evidence indicates that relieving inhibition by the TSC1/2 complex represents the mechanism through which leucine and insulin upregulate mTOR and downstream target activities (fig. 1). In particular, insulin may act through phosphorylation of TSC2 mediated by Akt/PKB whereas leucine seems to act through an alternative mechanism, which may involve inhibition of the association between mTOR and raptor [160].

An open question relative to stimulation of MPS mediated by BCAA is the potential contribution of insulin in the response. Neither physiological changes in circulating insulin levels nor administration of carbohydrates are sufficient to stimulate MPS in postabsorptive states [141, 155–157, 161–163]. In food-deprived rats, administration of a combination of carbohydrates with leucine stimulates MPS similarly to leucine alone [141, 161] and no difference in circulating insulin concentration arises between the two groups. Thus, leucine, unlike insulin, is the major determinant of the anabolic response in presence of food deprivation [141]. Anyhow, physiological circulating levels of insulin seem to be necessary for the anabolic response to BCAA [113, 144]. Indeed, a transient spike of circulating insulin has been found 15–45 min following leucine administration and may facilitate the stimulatory effect on MPS [164].

If, on the one hand, insulin seems to facilitate the role of leucine/BCAA as an 'anabolic hormone', on the other hand, amino acids are necessary for the insulin-induced activation of mTOR signaling and translation in both healthy, sarcopenic [D'Antona and Nisoli unpubl. results] and insulin-resistant skeletal muscle [165]. Overall, the role of insulin in mediating anabolic responses due to nutrients is of great importance when considering the failure in treating sarcopenia by means of a nutritional approach particularly in the presence of insulin resistance or diabetes.

The scenario arising from available studies indicates that the cross-talk between insulin and BCAA at the level of mTOR signaling may represent a crucial limiting factor of the physiological anabolic response to nutrients. Anyhow, of great importance, particularly in the presence of altered insulin sensitivity, is the possibility that BCAA may also act through insulin-independent and rapamycin-independent mechanisms of protein synthesis. As a matter of fact, the stimulatory effect of leucine on protein synthesis appears greater after administration to diabetic animals [113] and supplementation with BCAA-enriched mixture partially restores the metabolic impairment in heart [166] and skeletal muscle [167; D'Antona, unpubl. results] of an experimental model of type 1 diabetes, probably through an mTOR insulin-independent mechanism.

The known effects of amino acids on mRNA translation and MPS represent the rationale of the nutritional approach for treatment of muscular atrophies of different origin [168] and, in particular, of sarcopenia of aging. The knowledge of possible failures in signal transduction pathways in the elderly represents a fundamental prerequisite to optimize the therapeutic approach and facilitate the passage from 'what' to 'how' in prevention and treatment of sarcopenia in humans.

mTOR Signaling in Aging

In 1927, Ruger and Stoessiger [169] showed that in old age the skeletal muscle displays deterioration of function. Later, this feature was linked to a reduction of muscle mass.

Nowadays, the impact of aging on known pathways, i.e. insulin-IRS-PI3K-PDK1-Akt and mTOR-S6K1, involved in regulation of protein synthesis and steadiness of muscle mass is not completely explored and available data are partially contradictory.

First, conflicting results have been reported regarding MPS in aging. Several studies described a generalized reduction of resting MPS with age [170–177], while this observation has not been confirmed by others [178–180]. In particular, studies have shown that synthesis of mitochondrial proteins and myosin heavy chains (MHC) declines in aged muscles, as demonstrated by reduction in the availability of specific mRNAs [181], whereas synthesis of sarcoplasmic proteins may show a relative increase [170]. The reason for a selective age-related decrease of MPS remains unknown [182].

When present, a decline in resting MPS is attributed, at least in part, to decreased translational efficiency [181]. A significant reduction in the rate of protein synthesis in response to adequate stimuli of skeletal muscle growth, including mechanical load, electrical stimulation, insulin or amino acids, surely represents the major contributor to muscle dysfunction in the elderly. In fact, mechanically, electrically and nutritionally evoked increase in MPS, especially in fast-twitch muscle, are attenuated with old age [172, 183–185]. Despite data on the basal level of mTOR and S6K1 activation suggesting no change [46, 177, 186] or muscle specific increase [185] in aged in comparison with young skeletal muscle, the responsiveness of these signals to various stimuli is clearly impaired [151, 177, 185–187]. In particular, in a model of skeletal muscle compensatory overload, in which a significant impairment of MPS [184] and growth [188] has been reported in aged animals, Thomson and Gordon [189] have recently demonstrated age-related impairments in the phosphorylation states of mTOR (at Ser2448), S6K (at Thr389) and 4E-BP1 as well as deficits in total eEF2 accretion in fast twitch skeletal muscles. These findings strongly suggest that reduced overload-induced growth in aged muscles may be due partly to failure in activation of signals involved in the control of translation (fig. 1b).

Several studies have clearly shown a transient upregulation of components of the mTOR pathway, particularly following resistance training [190–192]. Moreover, the overall response appears reduced in the elderly [185, 193]. In particular, no change in MPS was observed in the old after a single bout of resistance exercise [194] as opposed to young subjects [195], and the response of the mTOR signaling appears significantly delayed and partially impaired [165]. Interestingly, no change in the same response has been described following intermittent passive stretch [196].

Importantly, alteration in MPS due to reduced translational efficiency in response to hormonal and nutritional stimuli, namely insulin and/or amino acids, in postprandial rather than in postabsorptive states, may represent an additional major mechanism underlying reduced net protein deposition in the elderly. It has been shown that whereas in postabsorptive conditions physiological changes in circulating insulin are not able to actively promote synthesis [144], in postprandial states the MPS appears upregulated. Moreover, in these phases the increased circulating insulin and amino acid availability represent the key components of the response. In fact, amino acids and, in particular, leucine have been shown to increase the sensitivity of MPS to insulin in vivo [150]. On the other hand, administration of leucine results in transient elevation in circulating concentration of insulin [113, 140, 179, 180]. Thus, the arising insulin resistance with age combined with inadequate amino acid stimulatory action may represent a limiting factor of MPS.

Many studies have outlined the impaired tolerance to glucose in the elderly [197–200], which begins in mid-life and worsens later. Reduced tolerance to glucose has been attributed, at least in part, to insulin resistance, as demonstrated by elevated plasma insulin levels [201]. Furthermore, the hyperglycamic/hyperinsulinemic clamp technique identified the peripheral (i.e. muscle) origin of the reduced sensitivity to glucose [197], downstream of IR [199]. Thus, the partial loss of the anticatabolic action of insulin, in postabsorptive states due to inhibition of muscle protein breakdown [202] and in postprandial states due to stimulation of synthesis [203], may impair MPS. This process might be exacerbated by reduced availability of amino acids in postprandial states and/or reduced MPS in response to BCAA in the elderly. In fact, in vitro assays demonstrated that the aging process reduces neutral amino acid transport across the cell membranes, thus reducing the intracellular neutral amino acids pools [204], and modifies the serum amino acid profile which appears different from that observed in the young [205]. Furthermore, the synthetic response to BCAA appears attenuated as the dose dependency curves of leucine-induced protein synthesis show a significant shift to the right. This change indicates the need for adequate amino acid supplementation in order to correct the MPS stimulation to the values observed in the young [151, 165, 206–208].

Notwithstanding that in aging and, possibly, in a bulk of other atrophic disorders, there is apparently reduced sensitivity to the stimulatory effects of EAA at physiological concentration, increased availability of amino acids by adequate supplementation can still stimulate MPS. Nevertheless, attempts to counteract sarcopenia of aging with nutritional supplements or increased protein intake have initially failed, at least as reflected by mass and torque recovery [209, 210].

There are several possible reasons contributing to the failure of supplements to counteract sarcopenia: reduced energy intake, combination, composition, duration, spread or pulse feeding and age. First the ineffectiveness of nutritional supplementation can be attributed to the observation that elderly subjects would voluntarily reduce their normal daily energy intake to compensate for the nutritional supplement [211]. In this case, the supplement simply acts as an energy supply and does not confer any additional nutritional benefit. Moreover, composition and combination of the supplement with other nutritional substrates may represent crucial determinants of the physical response. An amino acid supplement should elicit greater anabolism (guaranteed by presence of leucine) than normal feeding or corresponding energy intake and should contain EAA as non-EAA failed to have any direct effect on MPS [137, 138]. Ideally the EAA in the supplement should be represented in specific reciprocal proportions based on consumption in basal conditions [212]. In fact, among EAA, leucine, isoleucine, valine, lysine and threonine cover ~75% of the whole nitrogen need in basal conditions whereas histidine supplies the remaining 25% [213]. Furthermore, there is evidence that combining EAA with carbohydrates in a nutritional supplement is not as beneficial as the administration of EAA alone in promoting the anabolic response in the elderly [179]. Indeed, the reduced protein breakdown due to carbohydrates may prevent the anabolic response to EAA through impaired muscle remodeling [179]. Interestingly, the dairy distribution of supplementation may also significantly impact the outcome. In fact, more positive results have been obtained with pulse feeding (almost all dairy amino acids in a meal) than spread feeding (dairy amino acids equally distributed) [214, 215]. Finally, duration of treatment and age of intervention may represent additional critical factors for adequacy of MPS stimulation. The supply duration may cover the average fractional rate of muscular protein synthesis calculated at the age of intervention. For example in 22-month-old rats, in the presence of a daily rate of synthesis of about 5.7%/day [216], complete protein turnover theoretically occurs in at least 18 days and an effective supplementation may overcome this period of time. Finally, it is predictable that in the very elderly an abrupt reduction of the anabolic response to EAA occurs, thus preventing successful nutritional rescue of sarcopenia.

Although initial attempts in promoting positive effects with protein/amino acid supplements have failed and in the past decade a growing body of evidence in both humans and animals has revealed that protein intakes above the recommended daily allowance (RDA) are beneficial and useful in the treatment of diseases, including obesity [217], osteoporosis [218], type 2 diabetes (T2DM) [124, 125], chronic heart failure [219–221], heart disease [222–224], neurological disorders [225, 226], steroid myopathy [211], and sarcopenia [46, 47, 168, 227]. These outcomes may derive from cell signaling via leucine [228, 229], satiety [217, 230], amelioration in glycemic control [124, 125, 231, 232], and muscle fiber composition [233]. Notably, recent evidence suggests that prolonged administration of balanced EAA mixtures may represent a successful approach, leading to prevention of the age-related fiber atrophy, and to increased lean body mass and mitochondrial biogenesis, with improved muscle torque gain and muscle phenotype [46, 47, 168]. Preliminary findings suggest that these actions may derive, at least in part, from increased activation of mTOR signaling [46; D'Antona and Nisoli unpubl. obs.] and restoration of sensitivity of the mTOR pathway to insulin [D'Antona and Nisoli, unpubl. results].

mTOR, Mitochondrial Biogenesis, and Sarcopenia

The malfunction of mitochondria and the decrease of mitochondrial biogenesis, together with increased oxidative damage seem to exert some of the most deleterious effects of aging on different organs and tissues, including skeletal muscle [234].

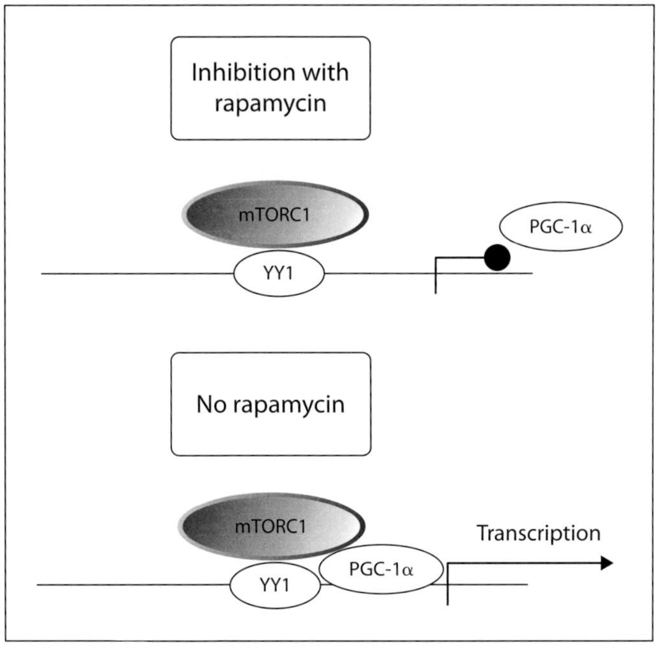

Fig. 2. Schematic depiction of rapamycin-dependent coactivation of mTORC1, PGC1-α and YY1 leading to transcriptional activity of mitochondrial genes. Modified from Cunningham et al. [242]. mTORC1 = Mammalian target of rapamycin complex 1; PGC1-α = peroxisone proliferator-activated receptor-γ coactivator 1α; YY1 = transcription factor yin-yang 1.

Apoptosis as well as autophagy have been implicated in disease mechanisms [235]. Therapeutic strategies for sarcopenia like endurance exercise [236] and calorie restriction [235] result in increased mitochondrial capacity in the muscle strongly supporting that mitochondrial dysfunction has a critical role in the muscle loss.

A key player controlling mitochondrial function is the peroxisome proliferator-activated receptor-γ coactivator-α (PGC-1α), a master regulator of mitochondrial biogenesis [237] (fig. 2). In skeletal muscle, PGC-1α can also prevent muscle wasting by regulating autophagy [238] and stabilization of the neuromuscular junction program [239] in the context of muscle atrophy during disease. Thus, PGC-1α links mitochondrial function to muscle integrity [238]. PGC-1α levels in skeletal muscle decrease during aging [240]. Recently, Wenz et al. [241] have analyzed the effect of mildly increased PGC-1α expression in skeletal muscle during aging. They found that muscle PGC-1α overexpression in MCKPGC-1α transgenic animals preserved mitochondrial function, neuromuscular junctions, and muscle integrity during aging. Increased PGC-1α levels in skeletal muscle prevented muscle wasting by reducing apoptosis, autophagy, and proteasome degradation. The preservation of muscle integrity and function in MCK-PGC-1α animals resulted in significantly improved whole-body health; both the loss of bone mineral density and the increase of systemic chronic inflammation, observed during normal aging, were prevented. Importantly, MCKPGC-1α animals also showed improved metabolic responses as evident by increased insulin sensitivity and insulin signaling in aged mice [241].

Recent evidence supports that mTOR is necessary for the maintenance of mitochondrial oxidative function. In skeletal muscle tissues and cells, the mTOR inhibitor rapamycin decreased the gene expression of the mitochondrial transcriptional regulators PGC-1a, estrogen-related receptor α, and nuclear respiratory factors (NRF), resulting in a decrease in mitochondrial gene expression and oxygen consumption [242]. Using computational genomics, the authors identified the transcription factor yin-yang 1 (YY1) as a common target of mTOR and PGC-1α. Knockdown of YY1 caused a significant decrease in mitochondrial gene expression and in respiration, and YY1 was required for rapamycin-dependent repression of those genes. Moreover, mTOR and raptor interacted with YY1, and inhibition of mTOR resulted in a failure of YY1 to interact with and be coactivated by PGC-1α [242]. The authors suggested this transcriptional mechanism as the means by which mTOR controls mitochondrial function; however, treatment with rapamycin does not result in decreased mitochondrial content even after 8 h [242].

More recently, Ramanathan and Schreiber [243] have described an immediate change in mitochondrial function following the inhibition of mTOR. Using global physiological profiling, the authors showed that inhibition of mTOR has immediate effects on carbon and mitochondrial metabolism in Jurkat cells, a leukemic cell line. Rapamycin-treated leukemic cells display reduced mitochondrial function, resulting in energy production via enhanced aerobic glycolysis in preference over mitochondrial respiration. It has been found that mTOR coimmunoprecipitates with the outer mitochondrial membrane proteins VDAC1 and Bcl-xl. Bcl-xl, a key mediator of mitochondrial function and cellular apoptosis has been shown to bind to VDAC1 and to increase substrate permeability [244]. VDAC proteins, which are situated in the outer mitochondrial membrane and are known to be key mediators of substrate transport into the mitochondria [245].

It has been previously demonstrated that mTOR colocalizes with the outer membrane of the mitochondria in Jurkat and 3T3-L1 adipocytes [246], suggesting that the direct control of mitochondrial function may also be present in other cell types besides leukemic cells. Accordingly, it would be of interest to investigate the mechanism by which mTOR might be targeted to the mitochondria in skeletal muscle cells. A member of the FK506-binding family of proteins, FKBP38, binds directly to the FRB domain of mTOR [247]. FKBP38 is localized on the mitochondria and has been shown to anchor proteins like Bcl2 and Bcl-xl to this organelle [248]. The FKBP38-mTOR interaction is not thought to be involved in the Rheb protein-mediated activation of mTOR [249]. Thus, mTOR might possibly be localized to the mitochondria via an FKBP38-mediated anchoring mechanism. The upstream signals to mitochondrial mTOR merit further investigation in muscle cells. mTOR has been demonstrated to be sensitive to mitochondrial dysfunction [246]. Localization of mTOR to the mitochondria might directly sensitize the activity of mTOR to the redox status of the cell. Another signal to mitochondrial mTOR, potentially relevant in muscle cells, might be through the AMP-dependent protein kinase, an important sensor of energy in cells.

AMPK is known to activate TSC2 (via the tumor suppressor LKB1) and to inhibit mTOR [250].

Notably, all of these cellular mechanisms ought to be investigated in age-related sarcopenic muscles. Thus, one can speculate that amino acids may be relevant as therapeutic approach to improve mitochondrial function in age-related sarcopenia by modulating mTOR signaling pathway.

References

1 Munro HN: Nutrition and muscle protein metabolism: Introduction. Fed Proc 1978;37:2281–2282.

2 Glass DJ: Signalling pathways that mediate skeletal muscle hypertrophy and atrophy. Nat Cell Biol 2003;5:87–90.

3 Bilmazes C, Uauy R, Haverberg LN, Munro HN, Young VR: Musle protein breakdown rates in humans based on Ntau-methylhistidine (3-methylhistidine) content of mixed proteins in skeletal muscle and urinary output of Ntau-methylhistidine. Metabolism 1978;27:525–530.

4 Halliday D, McKeran RO: Measurement of muscle protein synthetic rate from serial muscle biopsies and total body protein turnover in man by continuous intravenous infusion of L-(alpha-15N)lysine. Clin Sci Mol Med 1975;49:581–590.

5 McKeran RO, Halliday D, Purkiss P: Comparison of human myofibrillar protein catabolic rate derived from 3-methylhistidine excretion with synthetic rate from muscle biopsies during L-[alpha-15N] lysine infusion. Clin Sci Mol Med 1978;54:471–475.

6 Neuhauser M, Bassler KH: Endogenous 3-methylhistidine excretion in healthy women and men with reference to muscle protein metabolism. Z Ernahrungswiss 1984;23:171–180.

7 Elia M, Carter A, Bacon S, Winearls CG, Smith R: Clinical usefulness of urinary 3-methylhistidine excretion in indicating muscle protein breakdown. Br Med J 1981;282:351–354.

8 Tomas FM, Ballard FJ, Pope LM: Age-dependent changes in the rate of myofibrillar protein degradation in humans as assessed by 3-methylhistidine and creatinine excretion. Clin Sci (Lond) 1979;56:341–346.

9 Ballard FJ, Tomas FM, Stern LM: Increased turnover of muscle contractile proteins in Duchenne muscular dystrophy as assessed by 3-methylhistidine and creatinine excretion. Clin Sci (Lond) 1979;56:347–352.

10 Burgoyne JL, Ballard FJ, Tomas FM, Dobozy A, MacLennan AH, Fitzgerald A, Dahlenburg GW: Measurements of myofibrillar protein breakdown in newborn human infants. Clin Sci (Lond) 1982;63:421–427.

11 Siebrits FK, Barnes PM: The change in the rate of muscle protein metabolism of rats from weaning to 90 days of age. Comp Biochem Physiol A Comp Physiol 1989;92:485–488.

12 Dutta C, Hadley EC: The significance of sarcopenia in old age. J Gerontol A Biol Sci Med Sci 1995;50:1–4.

13 Baumgartner RN, Koehler KM, Gallagher D, Romero L, Heymsfield SB, Ross RR, Garry PJ, Lindeman RD: Epidemiology of sarcopenia among the elderly in New Mexico. Am J Epidemiol 1998;147:755–763.

14 D'Antona G, Pellegrino MA, Adami R, Rossi R, Carlizzi CN, Canepari M, Saltin B, Bottinelli R: The effect of ageing and immobilization on structure and function of human skeletal muscle fibres. J Physiol 2003;552:499–511.

15 D'Antona G, Pellegrino MA, Carlizzi CN, Bottinelli R: Deterioration of contractile properties of muscle fibres in elderly subjects is modulated by the level of physical activity. Eur J Appl Physiol 2007;100:603–611.

16 Fruhbeck G, Muguerza B, Castilla-Cortazar I, Santidrian S: Effect of aging on the rate of muscle protein turnover in rat. Rev Esp Fisiol 1996;52:207–214.

17 Dahlmann B, Schroeter C, Herbertz L, Reinauer H: Myofibrillar protein degradation and muscle proteinases in normal and diabetic rats. Biochem Med 1979;21:33–39.

18 Furuno K, Goodman MN, Goldberg AL: Role of different proteolytic systems in the degradation of muscle proteins during denervation atrophy. J Biol Chem 1990;265:8550–8557.

19 Warnes DM, Tomas FM, Ballard FJ: Increased rates of myofibrillar protein breakdown in muscle-wasting diseases. Muscle Nerve 1981;4:62–66.

20 McKeran RO, Halliday D, Purkiss P: Increased myofibrillar protein catabolism in Duchenne muscular dystrophy measured by 3-methylhistidine excretion in the urine. J Neurol Neurosurg Psychiatry 1977;40:979–981.
21 Hall-Angeras M, Angeras U, Zamir O, Hasselgren PO, Fischer JE: Effect of the glucocorticoid receptor antagonist RU 38486 on muscle protein breakdown in sepsis. Surgery 1991;109:468–473.
22 Hall-Angeras M, Hasselgren PO, Dimlich RV, Fischer JE: Myofibrillar proteinase, cathepsin B, and protein breakdown rates in skeletal muscle from septic rats. Metabolism 1991;40:302–306.
23 Hasselgren PO, James JH, Benson DW, Hall-Angeras M, Angeras U, Hiyama DT, Li S, Fischer JE: Total and myofibrillar protein breakdown in different types of rat skeletal muscle: effects of sepsis and regulation by insulin. Metabolism 1989;38:634–640.
24 Hasselgren PO, James JH, Warner BW, Hummel RP 3rd, Fischer JE: Protein synthesis and degradation in skeletal muscle from septic rats. Response to leucine and alpha-ketoisocaproic acid. Arch Surg 1988;123:640–644.
25 Demark-Wahnefried W, Peterson BL, Winer EP, Marks L, Aziz N, Marcom PK, Blackwell K, Rimer BK: Changes in weight, body composition, and factors influencing energy balance among premenopausal breast cancer patients receiving adjuvant chemotherapy. J Clin Oncol 2001;19:2381–2389.
26 Freedman RJ, Aziz N, Albanes D, Hartman T, Danforth D, Hill S, Sebring N, Reynolds JC, Yanovski JA: Weight and body composition changes during and after adjuvant chemotherapy in women with breast cancer. J Clin Endocrinol Metab 2004;89:2248–2253.
27 Penna F, Bonetto A, Muscaritoli M, Costamagna D, Minero VG, Bonelli G, Rossi Fanelli F, Baccino FM, Costelli P: Muscle atrophy in experimental cancer cachexia: is the IGF-1 signaling pathway involved? Int J Cancer 2010;in press.
28 Bolster DR, Jefferson LS, Kimball SR: Regulation of protein synthesis associated with skeletal muscle hypertrophy by insulin-, amino acid- and exercise-induced signalling. Proc Nutr Soc 2004;63:351–356.
29 Sandri M: Signaling in muscle atrophy and hypertrophy. Physiology 2008;23:160–170.
30 Latres E, Amini AR, Amini AA, Griffiths J, Martin FJ, Wei Y, Lin HC, Yancopoulos GD, Glass DJ: Insulin-like growth factor-1 (IGF-1) inversely regulates atrophy-induced genes via the phosphatidylinositol 3-kinase/Akt/mammalian target of rapamycin (PI3K/Akt/mTOR) pathway. J Biol Chem 2005;280:2737–2744.
31 Lecker SH, Jagoe RT, Gilbert A, Gomes M, Baracos V, Bailey J, Price SR, Mitch WE, Goldberg AL: Multiple types of skeletal muscle atrophy involve a common program of changes in gene expression. FASEB J 2004;18:39–51.
32 Jagoe RT, Lecker SH, Gomes M, Goldberg AL: Patterns of gene expression in atrophying skeletal muscles: response to food deprivation. FASEB J 2002;16:1697–1712.
33 Tawa NE Jr, Odessey R, Goldberg AL: Inhibitors of the proteasome reduce the accelerated proteolysis in atrophying rat skeletal muscles. J Clin Invest 1997;100:197–203.
34 Bodine SC, Latres E, Baumhueter S, Lai VK, Nunez L, Clarke BA, Poueymirou WT, Panaro FJ, Na E, Dharmarajan K, Pan ZQ, Valenzuela DM, DeChiara TM, Stitt TN, Yancopoulos GD, Glass DJ: Identification of ubiquitin ligases required for skeletal muscle atrophy. Science 2001;294:1704–1708.
35 Gomes MD, Lecker SH, Jagoe RT, Navon A, Goldberg AL: Atrogin-1, a muscle-specific F-box protein highly expressed during muscle atrophy. Proc Natl Acad Sci USA 2001;98:14440–14445.
36 Stevenson EJ, Giresi PG, Koncarevic A, Kandarian SC: Global analysis of gene expression patterns during disuse atrophy in rat skeletal muscle. J Physiol 2003;551:33–48.
37 Goldspink G: Changes in muscle mass and phenotype and the expression of autocrine and systemic growth factors by muscle in response to stretch and overload. J Anat 1999;194:323–334.
38 Musaro A, McCullagh K, Paul A, Houghton L, Dobrowolny G, Molinaro M, Barton ER, Sweeney HL, Rosenthal N: Localized Igf-1 transgene expression sustains hypertrophy and regeneration in senescent skeletal muscle. Nat Genet 2001;27:195–200.
39 Coleman ME, DeMayo F, Yin KC, Lee HM, Geske R, Montgomery C, Schwartz RJ: Myogenic vector expression of insulin-like growth factor I stimulates muscle cell differentiation and myofiber hypertrophy in transgenic mice. J Biol Chem 1995;270:12109–12116.
40 McCall GE, Allen DL, Haddad F, Baldwin KM: Transcriptional regulation of IGF-I expression in skeletal muscle. Am J Physiol Cell Physiol 2003;285:C831–C839.
41 DeVol DL, Rotwein P, Sadow JL, Novakofski J, Bechtel PJ: Activation of insulin-like growth factor gene expression during work-induced skeletal muscle growth. Am J Physiol 1990;259:E89–E95.
42 Lai KM, Gonzalez M, Poueymirou WT, Kline WO, Na E, Zlotchenko E, Stitt TN, Economides AN, Yancopoulos GD, Glass DJ: Conditional activation of akt in adult skeletal muscle induces rapid hypertrophy. Mol Cell Biol 2004;24:9295–9304.

43 Sugita H, Kaneki M, Sugita M, Yasukawa T, Yasuhara S, Martyn JA: Burn injury impairs insulin-stimulated Akt/PKB activation in skeletal muscle. Am J Physiol Endocrinol Metab 2005;288:E585–E591.
44 Sandri M, Sandri C, Gilbert A, Skurk C, Calabria E, Picard A, Walsh K, Schiaffino S, Lecker SH, Goldberg AL: Foxo transcription factors induce the atrophy-related ubiquitin ligase atrogin-1 and cause skeletal muscle atrophy. Cell 2004;117:399–412.
45 Stitt TN, Drujan D, Clarke BA, Panaro F, Timofeyva Y, Kline WO, Gonzalez M, Yancopoulos GD, Glass DJ: The IGF-1/PI3K/Akt pathway prevents expression of muscle atrophy-induced ubiquitin ligases by inhibiting FOXO transcription factors. Mol Cell 2004;14:395–403.
46 Pansarasa O, Flati V, Corsetti G, Brocca L, Pasini E, D'Antona G: Oral amino acid supplementation counteracts age-induced sarcopenia in elderly rats. Am J Cardiol 2008;101:35E–41E.
47 Corsetti G, Pasini E, D'Antona G, Nisoli E, Flati V, Assanelli D, Dioguardi FS, Bianchi R: Morphometric changes induced by amino acid supplementation in skeletal and cardiac muscles of old mice. Am J Cardiol 2008;101:26E–34E.
48 Parente V, D'Antona G, Adami R, Miotti D, Capodaglio P, De Vito G, Bottinelli R: Long-term resistance training improves force and unloaded shortening velocity of single muscle fibres of elderly women. Eur J Appl Physiol 2008;104:885–893.
49 Bohni R, Riesgo-Escovar J, Oldham S, Brogiolo W, Stocker H, Andruss BF, Beckingham K, Hafen E: Autonomous control of cell and organ size by CHICO, a *Drosophila* homolog of vertebrate IRS1–4. Cell 1999;97:865–875.
50 Huang H, Potter CJ, Tao W, Li DM, Brogiolo W, Hafen E, Sun H, Xu T: PTEN affects cell size, cell proliferation and apoptosis during *Drosophila* eye development. Development 1999;126:5365–5372.
51 Leevers SJ, Weinkove D, MacDougall LK, Hafen E, Waterfield MD: The Drosophila phosphoinositide 3-kinase Dp110 promotes cell growth. EMBO J 1996;15:6584–6594.
52 Montagne J, Stewart MJ, Stocker H, Hafen E, Kozma SC, Thomas G: *Drosophila* S6 kinase: a regulator of cell size. Science 1999;285:2126–2129.
53 Datta SR, Brunet A, Greenberg ME: Cellular survival: a play in three Akts. Genes Dev 1999;13:2905–2927.
54 Nicholson KM, Anderson NG: The protein kinase B/Akt signalling pathway in human malignancy. Cell Signal 2002;14:381–395.
55 Jones PF, Jakubowicz T, Pitossi FJ, Maurer F, Hemmings BA: Molecular cloning and identification of a serine/threonine protein kinase of the second-messenger subfamily. Proc Natl Acad Sci USA 1991;88:4171–4175.
56 Andjelkovic M, Jones PF, Grossniklaus U, Cron P, Schier AF, Dick M, Bilbe G, Hemmings BA: Developmental regulation of expression and activity of multiple forms of the *Drosophila* RAC protein kinase. J Biol Chem 1995;270:4066–4075.
57 Alessi DR, Andjelkovic M, Caudwell B, Cron P, Morrice N, Cohen P, Hemmings BA: Mechanism of activation of protein kinase B by insulin and IGF-1. EMBO J 1996;15:6541–6551.
58 Burgering BM, Coffer PJ: Protein kinase B (c-Akt) in phosphatidylinositol-3-OH kinase signal transduction. Nature 1995;376:599–602.
59 Franke TF, Yang SI, Chan TO, Datta K, Kazlauskas A, Morrison DK, Kaplan DR, Tsichlis PN: The protein kinase encoded by the Akt proto-oncogene is a target of the PDGF-activated phosphatidylinositol 3-kinase. Cell 1995;81:727–736.
60 Bodine SC, Stitt TN, Gonzalez M, Kline WO, Stover GL, Bauerlein R, Zlotchenko E, Scrimgeour A, Lawrence JC, Glass DJ, Yancopoulos GD: Akt/mTOR pathway is a crucial regulator of skeletal muscle hypertrophy and can prevent muscle atrophy in vivo. Nat Cell Biol 2001;3:1014–1019.
61 Pallafacchina G, Calabria E, Serrano AL, Kalhovde JM, Schiaffino S: A protein kinase B-dependent and rapamycin-sensitive pathway controls skeletal muscle growth but not fiber type specification. Proc Natl Acad Sci USA 2002;99:9213–9218.
62 Nader GA, Esser KA: Intracellular signaling specificity in skeletal muscle in response to different modes of exercise. J Appl Physiol 2001;90:1936–1942.
63 Sakamoto K, Aschenbach WG, Hirshman MF, Goodyear LJ: Akt signaling in skeletal muscle: regulation by exercise and passive stretch. Am J Physiol Endocrinol Metab 2003;285:E1081–E1088.
64 Sakamoto K, Arnolds DE, Ekberg I, Thorell A, Goodyear LJ: Exercise regulates Akt and glycogen synthase kinase-3 activities in human skeletal muscle. Biochem Biophys Res Commun 2004;319:419–425.
65 Rommel C, Bodine SC, Clarke BA, Rossman R, Nunez L, Stitt TN, Yancopoulos GD, Glass DJ: Mediation of IGF-1-induced skeletal myotube hypertrophy by PI(3)K/Akt/mTOR and PI(3)K/Akt/GSK3 pathways. Nat Cell Biol 2001;3:1009–1013.
66 Vivanco I, Sawyers CL: The phosphatidylinositol 3-kinase AKT pathway in human cancer. Nat Rev Cancer 2002;2:489–501.
67 Sarbassov DD, Guertin DA, Ali SM, Sabatini DM: Phosphorylation and regulation of Akt/PKB by the rictor-mTOR complex. Science 2005;307:1098–1101.

68 Wang X, Proud CG: The mTOR pathway in the control of protein synthesis. Physiology (Bethesda) 2006;21:362–369.
69 Crespo JL, Powers T, Fowler B, Hall MN: The TOR-controlled transcription activators GLN3, RTG1, and RTG3 are regulated in response to intracellular levels of glutamine. Proc Natl Acad Sci USA 2002;99: 6784–6789.
70 Crespo JL, Diaz-Troya S, Florencio FJ: Inhibition of target of rapamycin signaling by rapamycin in the unicellular green alga *Chlamydomonas reinhardtii*. Plant Physiol 2005;139:1736–1749.
71 Huang S, Bjornsti MA, Houghton PJ: Rapamycins: mechanism of action and cellular resistance. Cancer Biol Ther 2003;2:222–232.
72 Sarbassov DD, Ali SM, Sabatini DM: Growing roles for the mTOR pathway. Curr Opin Cell Biol 2005; 17:596–603.
73 Wullschleger S, Loewith R, Hall MN: TOR signaling in growth and metabolism. Cell 2006;124:471–484.
74 Avruch J: Insulin signal transduction through protein kinase cascades. Mol Cell Biochem 1998;182: 31–48.
75 Kozma SC, Thomas G: p70s6k/p85s6k: mechanism of activation and role in mitogenesis. Semin Cancer Biol 1994;5:255–260.
76 Vary TC, Lynch CJ: Nutrient signaling components controlling protein synthesis in striated muscle. J Nutr 2007;137:1835–1843.
77 Rhoads RE, Joshi B, Minich WB: Participation of initiation factors in the recruitment of mRNA to ribosomes. Biochimie 1994;76:831–838.
78 Rhoads RE: Regulation of eukaryotic protein synthesis by initiation factors. J Biol Chem 1993;268: 3017–3020.
79 Choi KM, McMahon LP, Lawrence JC Jr: Two motifs in the translational repressor PHAS-I required for efficient phosphorylation by mammalian target of rapamycin and for recognition by raptor. J Biol Chem 2003;278:19667–19673.
80 Flati V, Pasini E, D'Antona G, Speca S, Toniato E, Martinotti S: Intracellular mechanisms of metabolism regulation: the role of signaling via the mammalian target of rapamycin pathway and other routes. Am J Cardiol 2008;101:16E–21E.
81 Haghighat A, Mader S, Pause A, Sonenberg N: Repression of cap-dependent translation by 4E-binding protein 1: competition with p220 for binding to eukaryotic initiation factor-4E. EMBO J 1995;14:5701–5709.
82 Harris TE, Lawrence JC Jr: TOR signaling. Sci STKE 2003;2003:re15.
83 Ohanna M, Sobering AK, Lapointe T, Lorenzo L, Praud C, Petroulakis E, Sonenberg N, Kelly PA, Sotiropoulos A, Pende M: Atrophy of S6K1(–/–) skeletal muscle cells reveals distinct mTOR effectors for cell cycle and size control. Nat Cell Biol 2005; 7:286–294.
84 Wang X, Li W, Williams M, Terada N, Alessi DR, Proud CG: Regulation of elongation factor 2 kinase by p90(RSK1) and p70 S6 kinase. EMBO J 2001; 20:4370–4379.
85 Redpath NT, Foulstone EJ, Proud CG: Regulation of translation elongation factor-2 by insulin via a rapamycin-sensitive signalling pathway. EMBO J 1996; 15:2291–2297.
86 Hannan KM, Thomas G, Pearson RB: Activation of S6K1 (p70 ribosomal protein S6 kinase 1) requires an initial calcium-dependent priming event involving formation of a high-molecular-mass signalling complex. Biochem J 2003;370:469–477.
87 Glass DJ: Skeletal muscle hypertrophy and atrophy signaling pathways. Int J Biochem Cell Biol 2005;37: 1974–1984.
88 Pullen N, Dennis PB, Andjelkovic M, Dufner A, Kozma SC, Hemmings BA, Thomas G: Phosphorylation and activation of p70s6k by PDK1. Science 1998;279:707–710.
89 Hornberger TA, Hunter RB, Kandarian SC, Esser KA: Regulation of translation factors during hindlimb unloading and denervation of skeletal muscle in rats. Am J Physiol Cell Physiol 2001;281: C179–C187.
90 Chang L, Chiang SH, Saltiel AR: Insulin signaling and the regulation of glucose transport. Mol Med 2004;10:65–71.
91 Holz MK, Blenis J: Identification of S6 kinase 1 as a novel mammalian target of rapamycin (mTOR)-phosphorylating kinase. J Biol Chem 2005;280: 26089–26093.
92 Watson RT, Kanzaki M, Pessin JE: Regulated membrane trafficking of the insulin-responsive glucose transporter 4 in adipocytes. Endocr Rev 2004;25: 177–204.
93 Saltiel AR, Pessin JE: Insulin signaling in microdomains of the plasma membrane. Traffic 2003;4:711–716.
94 Maehama T, Dixon JE: PTEN: a tumour suppressor that functions as a phospholipid phosphatase. Trends Cell Biol 1999;9:125–128.
95 Pesesse X, Deleu S, De Smedt F, Drayer L, Erneux C: Identification of a second SH2-domain-containing protein closely related to the phosphatidylinositol polyphosphate 5-phosphatase SHIP. Biochem Biophys Res Commun 1997;239:697–700.

96 Sun XJ, Rothenberg P, Kahn CR, Backer JM, Araki E, Wilden PA, Cahill DA, Goldstein BJ, White MF: Structure of the insulin receptor substrate IRS-1 defines a unique signal transduction protein. Nature 1991;352:73–77.
97 Kido Y, Nakae J, Accili D: Clinical review 125:The insulin receptor and its cellular targets. J Clin Endocrinol Metab 2001;86:972–979.
98 Kohn AD, Barthel A, Kovacina KS, Boge A, Wallach B, Summers SA, Birnbaum MJ, Scott PH, Lawrence JC Jr, Roth RA: Construction and characterization of a conditionally active version of the serine/threonine kinase Akt. J Biol Chem 1998;273:11937–11943.
99 Corvera S, Czech MP: Direct targets of phosphoinositide 3-kinase products in membrane traffic and signal transduction. Trends Cell Biol 1998;8:442–446.
100 van Slegtenhorst M, Nellist M, Nagelkerken B, Cheadle J, Snell R, van den Ouweland A, Reuser A, Sampson J, Halley D, van der Sluijs P: Interaction between hamartin and tuberin, the TSC1 and TSC2 gene products. Hum Mol Genet 1998;7:1053–1057.
101 Zhang Y, Gao X, Saucedo LJ, Ru B, Edgar BA, Pan D: Rheb is a direct target of the tuberous sclerosis tumour suppressor proteins: Nat Cell Biol 2003;5: 578–581.
102 Tee AR, Fingar DC, Manning BD, Kwiatkowski DJ, Cantley LC, Blenis J: Tuberous sclerosis complex-1 and -2 gene products function together to inhibit mammalian target of rapamycin (mTOR)-mediated downstream signaling. Proc Natl Acad Sci USA 2002;99:13571–13576.
103 Inoki K, Li Y, Xu T, Guan KL: Rheb GTPase is a direct target of TSC2 GAP activity and regulates mTOR signaling. Genes Dev 2003;17:1829–1834.
104 Potter CJ, Pedraza LG, Xu T: Akt regulates growth by directly phosphorylating Tsc2. Nat Cell Biol 2002;4:658–665.
105 Roux PP, Ballif BA, Anjum R, Gygi SP, Blenis J: Tumor-promoting phorbol esters and activated Ras inactivate the tuberous sclerosis tumor suppressor complex via p90 ribosomal S6 kinase. Proc Natl Acad Sci USA 2004;101:13489–13494.
106 Ma L, Chen Z, Erdjument-Bromage H, Tempst P, Pandolfi PP: Phosphorylation and functional inactivation of TSC2 by Erk implications for tuberous sclerosis and cancer pathogenesis: Cell 2005;121: 179–193.
107 Li Y, Inoki K, Vacratsis P, Guan KL: The p38 and MK2 kinase cascade phosphorylates tuberin, the tuberous sclerosis 2 gene product, and enhances its interaction with 14–3–3. J Biol Chem 2003;278: 13663–13671.
108 Nobukini T, Thomas G: The mTOR/S6K signalling pathway: the role of the TSC1/2 tumour suppressor complex and the proto-oncogene Rheb. Novartis Found Symp 2004;262:148–154; discussion 154–159, 265–268.
109 Garami A, Zwartkruis FJ, Nobukuni T, Joaquin M, Roccio M, Stocker H, Kozma SC, Hafen E, Bos JL, Thomas G: Insulin activation of Rheb, a mediator of mTOR/S6K/4E-BP signaling, is inhibited by TSC1 and 2. Mol Cell 2003;11:1457–1466.
110 Shah OJ, Wang Z, Hunter T: Inappropriate activation of the TSC/Rheb/mTOR/S6K cassette induces IRS1/2 depletion, insulin resistance, and cell survival deficiencies. Curr Biol 2004;14:1650–1656.
111 Grzelkowska K, Dardevet D, Balage M, Grizard J: Involvement of the rapamycin-sensitive pathway in the insulin regulation of muscle protein synthesis in streptozotocin-diabetic rats. J Endocrinol 1999;160: 137–145.
112 Kimball SR, Jefferson LS, Fadden P, Haystead TA, Lawrence JC Jr: Insulin and diabetes cause reciprocal changes in the association of eIF-4E and PHAS-I in rat skeletal muscle. Am J Physiol 1996;270: C705–C709.
113 Anthony JC, Reiter AK, Anthony TG, Crozier SJ, Lang CH, MacLean DA, Kimball SR, Jefferson LS: Orally administered leucine enhances protein synthesis in skeletal muscle of diabetic rats in the absence of increases in 4E-BP1 or S6K1 phosphorylation. Diabetes 2002;51:928–936.
114 Gual P, Le Marchand-Brustel Y, Tanti JF: Positive and negative regulation of insulin signaling through IRS-1 phosphorylation. Biochimie 2005;87:99–109.
115 Tremblay F, Jacques H, Marette A: Modulation of insulin action by dietary proteins and amino acids: role of the mammalian target of rapamycin nutrient sensing pathway. Curr Opin Clin Nutr Metab Care 2005;8:457–462.
116 Gual P, Gremeaux T, Gonzalez T, Le Marchand-Brustel Y, Tanti JF: MAP kinases and mTOR mediate insulin-induced phosphorylation of insulin receptor substrate-1 on serine residues 307, 612 and 632. Diabetologia 2003;46:1532–1542.
117 Werner AB, Tait SW, de Vries E, Eldering E, Borst J: Requirement for aspartate-cleaved bid in apoptosis signaling by DNA-damaging anti-cancer regimens. J Biol Chem 2004;279:28771–28780.
118 Robinson KA, Buse MG: Mechanisms of high-glucose/insulin-mediated desensitization of acute insulin-stimulated glucose transport and Akt activation. Am J Physiol Endocrinol Metab 2008;294: E870–E881.

119 Hiratani K, Haruta T, Tani A, Kawahara J, Usui I, Kobayashi M: Roles of mTOR and JNK in serine phosphorylation, translocation, and degradation of IRS-1. Biochem Biophys Res Commun 2005;335: 836–842.
120 Um SH, Frigerio F, Watanabe M, Picard F, Joaquin M, Sticker M, Fumagalli S, Allegrini PR, Kozma SC, Auwerx J, Thomas G: Absence of S6K1 protects against age- and diet-induced obesity while enhancing insulin sensitivity. Nature 2004;431:200–205.
121 Hillier T, Long W, Jahn L, Wei L, Barrett EJ: Physiological hyperinsulinemia stimulates p70(S6k) phosphorylation in human skeletal muscle. J Clin Endocrinol Metab 2000;85:4900–4904.
122 Tremblay F, Marette A: Amino acid and insulin signaling via the mTOR/p70 S6 kinase pathway: a negative feedback mechanism leading to insulin resistance in skeletal muscle cells. J Biol Chem 2001; 276:38052–38060.
123 Patti ME, Brambilla E, Luzi L, Landaker EJ, Kahn CR: Bidirectional modulation of insulin action by amino acids. J Clin Invest 1998;101:1519–1529.
124 Solerte SB, Fioravanti M, Locatelli E, Bonacasa R, Zamboni M, Basso C, Mazzoleni A, Mansi V, Geroutis N, Gazzaruso C: Improvement of blood glucose control and insulin sensitivity during a long-term (60 weeks) randomized study with amino acid dietary supplements in elderly subjects with type 2 diabetes mellitus. Am J Cardiol 2008;101:82E–88E.
125 Solerte SB, Gazzaruso C, Bonacasa R, Rondanelli M, Zamboni M, Basso C, Locatelli E, Schifino N, Giustina A, Fioravanti M: Nutritional supplements with oral amino acid mixtures increases whole-body lean mass and insulin sensitivity in elderly subjects with sarcopenia. Am J Cardiol 2008;101:69E–77E.
126 Allison DJ, Clay T: Angiotensin II release from the hypoxic lobe of the intact dog lung (proceedings). J Physiol 1976;260:32P–33P.
127 Iritani N, Sugimoto T, Fukuda H, Komiya M, Ikeda H: Dietary soybean protein increases insulin receptor gene expression in Wistar fatty rats when dietary polyunsaturated fatty acid level is low. J Nutr 1997; 127:1077–1083.
128 Rice KM, Turnbow MA, Garner CW: Insulin stimulates the degradation of IRS-1 in 3T3-L1 adipocytes. Biochem Biophys Res Commun 1993;190:961–967.
129 Sun XJ, Goldberg JL, Qiao LY, Mitchell JJ: Insulin-induced insulin receptor substrate-1 degradation is mediated by the proteasome degradation pathway. Diabetes 1999;48:1359–1364.
130 Pederson TM, Kramer DL, Rondinone CM: Serine/threonine phosphorylation of IRS-1 triggers its degradation: possible regulation by tyrosine phosphorylation. Diabetes 2001;50:24–31.
131 Usui I, Imamura T, Huang J, Satoh H, Shenoy SK, Lefkowitz RJ, Hupfeld CJ, Olefsky JM: beta-arrestin-1 competitively inhibits insulin-induced ubiquitination and degradation of insulin receptor substrate 1. Mol Cell Biol 2004;24:8929–8937.
132 Ueki K, Kondo T, Kahn CR: Suppressor of cytokine signaling 1 (SOCS-1) and SOCS-3 cause insulin resistance through inhibition of tyrosine phosphorylation of insulin receptor substrate proteins by discrete mechanisms. Mol Cell Biol 2004;24:5434–5446.
133 Rakesh K, Agrawal DK: Controlling cytokine signaling by constitutive inhibitors. Biochem Pharmacol 2005;70:649–657.
134 Jasmin JF, Mercier I, Sotgia F, Lisanti MP: SOCS proteins and caveolin-1 as negative regulators of endocrine signaling. Trends Endocrinol Metab 2006;17:150–158.
135 Yoshimura A: Negative regulation of cytokine signaling. Clin Rev Allergy Immunol 2005;28:205–220.
136 Pasini E, Aquilani R, Dioguardi FS, D'Antona G, Gheorghiade M, Taegtmeyer H: Hypercatabolic syndrome: molecular basis and effects of nutritional supplements with amino acids. Am J Cardiol 2008; 101:11E–15E.
137 Smith K, Reynolds N, Downie S, Patel A, Rennie MJ: Effects of flooding amino acids on incorporation of labeled amino acids into human muscle protein. Am J Physiol 1998;275:E73–E78.
138 Smith K, Barua JM, Watt PW, Scrimgeour CM, Rennie MJ: Flooding with L-[$^{1-13}$C]leucine stimulates human muscle protein incorporation of continuously infused L-[$^{1-13}$C]valine. Am J Physiol 1992;262:E372–E376.
139 Biolo G, Tipton KD, Klein S, Wolfe RR: An abundant supply of amino acids enhances the metabolic effect of exercise on muscle protein. Am J Physiol 1997;273:E122–E129.
140 Volpi E, Ferrando AA, Yeckel CW, Tipton KD, Wolfe RR: Exogenous amino acids stimulate net muscle protein synthesis in the elderly. J Clin Invest 1998; 101:2000–2007.
141 Anthony JC, Anthony TG, Kimball SR, Vary TC, Jefferson LS: Orally administered leucine stimulates protein synthesis in skeletal muscle of postabsorptive rats in association with increased eIF4F formation. J Nutr 2000;130:139–145.
142 Svanberg E, Moller-Loswick AC, Matthews DE, Korner U, Andersson M, Lundholm K: Effects of amino acids on synthesis and degradation of skeletal muscle proteins in humans. Am J Physiol 1996; 271:E718–E724.
143 Svanberg E, Zachrisson H, Ohlsson C, Iresjo BM, Lundholm KG: Role of insulin and IGF-I in activation of muscle protein synthesis after oral feeding. Am J Physiol 1996;270:E614–E620.

144 Preedy VR, Garlick PJ: The response of muscle protein synthesis to nutrient intake in postabsorptive rats: the role of insulin and amino acids. Biosci Rep 1986;6:177–183.

145 Mordier S, Deval C, Bechet D, Tassa A, Ferrara M: Leucine limitation induces autophagy and activation of lysosome-dependent proteolysis in C2C12 myotubes through a mammalian target of rapamycin-independent signaling pathway. J Biol Chem 2000;275:29900–29906.

146 Bolster DR, Vary TC, Kimball SR, Jefferson LS: Leucine regulates translation initiation in rat skeletal muscle via enhanced eIF4G phosphorylation. J Nutr 2004;134:1704–1710.

147 Malaisse W: Branched chain amino acids and keto acids as regulators of insulin and glucagone release; in Adibi SA, Feki W, Langenbeck U, Schauder P (eds): Branched Chain Amino Acids and Keto Acids in Health and Disease. Basel, Karger, 1984, pp 119–133.

148 Lynch CJ, Halle B, Fujii H, Vary TC, Wallin R, Damuni Z, Hutson SM: Potential role of leucine metabolism in the leucine-signaling pathway involving mTOR. Am J Physiol Endocrinol Metab 2003; 285:E854–E863.

149 Lynch CJ, Patson BJ, Anthony J, Vaval A, Jefferson LS, Vary TC: Leucine is a direct-acting nutrient signal that regulates protein synthesis in adipose tissue. Am J Physiol Endocrinol Metab 2002;283: E503–E513.

150 Garlick PJ, Grant I: Amino acid infusion increases the sensitivity of muscle protein synthesis in vivo to insulin: effect of branched-chain amino acids. Biochem J 1988;254:579–584.

151 Dardevet D, Sornet C, Balage M, Grizard J: Stimulation of in vitro rat muscle protein synthesis by leucine decreases with age. J Nutr 2000;130:2630–2635.

152 Buse MG, Reid SS: Leucine. A possible regulator of protein turnover in muscle. J Clin Invest 1975;56: 1250–1261.

153 Li JB, Jefferson LS: Influence of amino acid availability on protein turnover in perfused skeletal muscle. Biochim Biophys Acta 1978;544:351–359.

154 McNurlan MA, Fern EB, Garlick PJ: Failure of leucine to stimulate protein synthesis in vivo. Biochem J 1982;204:831–838.

155 Yoshizawa F, Kimball SR, Jefferson LS: Modulation of translation initiation in rat skeletal muscle and liver in response to food intake. Biochem Biophys Res Commun 1997;240:825–831.

156 Yoshizawa F, Tonouchi A, Miura Y, Yagasaki K, Funabiki R: Insulin-stimulated polypeptide chain elongation in the soleus muscle of mice. Biosci Biotechnol Biochem 1995;59:348–349.

157 Yoshizawa F, Kimball SR, Vary TC, Jefferson LS: Effect of dietary protein on translation initiation in rat skeletal muscle and liver. Am J Physiol 1998; 275:E814–E820.

158 Anthony TG, Anthony JC, Yoshizawa F, Kimball SR, Jefferson LS: Oral administration of leucine stimulates ribosomal protein mRNA translation but not global rates of protein synthesis in the liver of rats. J Nutr 2001;131:1171–1176.

159 Schmidt A, Beck T, Koller A, Kunz J, Hall MN: The TOR nutrient signalling pathway phosphorylates NPR1 and inhibits turnover of the tryptophan permease. EMBO J 1998;17:6924–6931.

160 Kimball SR, Jefferson LS: Amino acids as regulators of gene expression. Nutr Metab (Lond) 2004;1:3.

161 Anthony JC, Anthony TG, Layman DK: Leucine supplementation enhances skeletal muscle recovery in rats following exercise. J Nutr 1999;129:1102–1106.

162 Gautsch TA, Anthony JC, Kimball SR, Paul GL, Layman DK, Jefferson LS: Availability of eIF4E regulates skeletal muscle protein synthesis during recovery from exercise. Am J Physiol 1998;274: C406–C414.

163 Garlick PJ, Fern M, Preedy VR: The effect of insulin infusion and food intake on muscle protein synthesis in postabsorptive rats. Biochem J 1983;210:669–676.

164 Anthony JC, Anthony TG, Kimball SR, Jefferson LS: Signaling pathways involved in translational control of protein synthesis in skeletal muscle by leucine. J Nutr 2001;131:856S–860S.

165 Drummond MJ, Dreyer HC, Pennings B, Fry CS, Dhanani S, Dillon EL, Sheffield-Moore M, Volpi E, Rasmussen BB: Skeletal muscle protein anabolic response to resistance exercise and essential amino acids is delayed with aging. J Appl Physiol 2008;104: 1452–1461.

166 Pellegrino MA, Patrini C, Pasini E, Brocca L, Flati V, Corsetti G, D'Antona G: Amino acid supplementation counteracts metabolic and functional damage in the diabetic rat heart. Am J Cardiol 2008;101:49E–56E.

167 Brocca L, D'Antona G, Bachi A, Pellegrino MA: Amino acid supplements improve native antioxidant enzyme expression in the skeletal muscle of diabetic mice. Am J Cardiol 2008;101:57E–62E.

168 Dillon EL, Sheffield-Moore M, Paddon-Jones D, Gilkison C, Sanford AP, Casperson SL, Jiang J, Chinkes DL, Urban RJ: Amino acid supplementation increases lean body mass, basal muscle protein synthesis, and insulin-like growth factor-I expression in older women. J Clin Endocrinol Metab 2009; 94:1630–1637.

169 Ruger HA, Stoessiger B: On the growth curves of certain characteristics in man (males). Ann Eugenics 1927;2:76–111.

170 Balagopal P, Rooyackers OE, Adey DB, Ades PA, Nair KS: Effects of aging on in vivo synthesis of skeletal muscle myosin heavy-chain and sarcoplasmic protein in humans. Am J Physiol 1997;273:E790–E800.

171 Welle S, Thornton C, Jozefowicz R, Statt M: Myofibrillar protein synthesis in young and old men. Am J Physiol 1993;264:E693–E698.

172 Welle S, Thornton C, Statt M: Myofibrillar protein synthesis in young and old human subjects after three months of resistance training. Am J Physiol 1995;268:E422–E427.

173 Welle S, Thornton C, Statt M, McHenry B: Postprandial myofibrillar and whole body protein synthesis in young and old human subjects. Am J Physiol 1994;267:E599–E604.

174 Yarasheski KE, Zachwieja JJ, Bier DM: Acute effects of resistance exercise on muscle protein synthesis rate in young and elderly men and women. Am J Physiol 1993;265:E210–E214.

175 Short KR, Vittone JL, Bigelow ML, Proctor DN, Nair KS: Age and aerobic exercise training effects on whole body and muscle protein metabolism. Am J Physiol Endocrinol Metab 2004;286:E92–E101.

176 Yarasheski KE, Welle S, Nair KS: Muscle protein synthesis in younger and older men. JAMA 2002;287:317–318.

177 Cuthbertson D, Smith K, Babraj J, Leese G, Waddell T, Atherton P, Wackerhage H, Taylor PM, Rennie MJ: Anabolic signaling deficits underlie amino acid resistance of wasting, aging muscle. FASEB J 2005;19:422–424.

178 Yarasheski KE, Pak-Loduca J, Hasten DL, Obert KA, Brown MB, Sinacore DR: Resistance exercise training increases mixed muscle protein synthesis rate in frail women and men >/ = 76 yr old. Am J Physiol 1999;277:E118–E125.

179 Volpi E, Mittendorfer B, Rasmussen BB, Wolfe RR: The response of muscle protein anabolism to combined hyperaminoacidemia and glucose-induced hyperinsulinemia is impaired in the elderly. J Clin Endocrinol Metab 2000;85:4481–4490.

180 Volpi E, Mittendorfer B, Wolf SE, Wolfe RR: Oral amino acids stimulate muscle protein anabolism in the elderly despite higher first-pass splanchnic extraction. Am J Physiol 1999;277:E513–E520.

181 Welle S, Bhatt K, Thornton C: Polyadenylated RNA, actin mRNA, and myosin heavy chain mRNA in young and old human skeletal muscle. Am J Physiol 1996;270:E224–E229.

182 Nair KS: Aging muscle. Am J Clin Nutr 2005;81:953–963.

183 Tamaki T, Uchiyama S, Uchiyama Y, Akatsuka A, Yoshimura S, Roy RR, Edgerton VR: Limited myogenic response to a single bout of weight-lifting exercise in old rats. Am J Physiol Cell Physiol 2000;278:C1143–C1152.

184 Pehme A, Alev K, Kaasik P, Seene T: Age-related changes in skeletal-muscle myosin heavy-chain composition: effect of mechanical loading. J Aging Phys Act 2004;12:29–44.

185 Parkington JD, LeBrasseur NK, Siebert AP, Fielding RA: Contraction-mediated mTOR, p70S6k, and ERK1/2 phosphorylation in aged skeletal muscle. J Appl Physiol 2004;97:243–248.

186 Li M, Li C, Parkhouse WS: Age-related differences in the des IGF-I-mediated activation of Akt-1 and p70 S6K in mouse skeletal muscle. Mech Age Dev 2003;124:771–778.

187 Guillet C, Prod'homme M, Balage M, Gachon P, Giraudet C, Morin L, Grizard J, Boirie Y: Impaired anabolic response of muscle protein synthesis is associated with S6K1 dysregulation in elderly humans. FASEB J 2004;18:1586–1587.

188 Thomson DM, Gordon SE: Diminished overload-induced hypertrophy in aged fast-twitch skeletal muscle is associated with AMPK hyperphosphorylation. J Appl Physiol 2005;98:557–564.

189 Thomson DM, Gordon SE: Impaired overload-induced muscle growth is associated with diminished translational signalling in aged rat fast-twitch skeletal muscle. J Physiol 2006;574:291–305.

190 Kubica N, Bolster DR, Farrell PA, Kimball SR, Jefferson LS: Resistance exercise increases muscle protein synthesis and translation of eukaryotic initiation factor 2Bepsilon mRNA in a mammalian target of rapamycin-dependent manner. J Biol Chem 2005;280:7570–7580.

191 Dreyer HC, Drummond MJ, Pennings B, Fujita S, Glynn EL, Chinkes DL, Dhanani S, Volpi E, Rasmussen BB: Leucine-enriched essential amino acid and carbohydrate ingestion following resistance exercise enhances mTOR signaling and protein synthesis in human muscle. Am J Physiol Endocrinol Metab 2008;294:E392–E400.

192 Bolster DR, Kubica N, Crozier SJ, Williamson DL, Farrell PA, Kimball SR, Jefferson LS: Immediate response of mammalian target of rapamycin (mTOR)-mediated signalling following acute resistance exercise in rat skeletal muscle. J Physiol 2003;553:213–220.

193 Funai K, Parkington JD, Carambula S, Fielding RA: Age-associated decrease in contraction-induced activation of downstream targets of Akt/mTor signaling in skeletal muscle. Am J Physiol Regul Integr Comp Physiol 2006;290:R1080–R1086.

194 Sheffield-Moore M, Paddon-Jones D, Sanford AP, Rosenblatt JI, Matlock AG, Cree MG, Wolfe RR: Mixed muscle and hepatic derived plasma protein metabolism is differentially regulated in older and younger men following resistance exercise. Am J Physiol Endocrinol Metab 2005;288:E922–E929.
195 Dreyer HC, Fujita S, Cadenas JG, Chinkes DL, Volpi E, Rasmussen BB: Resistance exercise increases AMPK activity and reduces 4E-BP1 phosphorylation and protein synthesis in human skeletal muscle. J Physiol 2006;576:613–624.
196 Hornberger TA, Mateja RD, Chin ER, Andrews JL, Esser KA: Aging does not alter the mechanosensitivity of the p38, p70S6k, and JNK2 signaling pathways in skeletal muscle. J Appl Physiol 2005;98:1562–1566.
197 DeFronzo RA, Hendler R, Simonson D: Insulin resistance is a prominent feature of insulin-dependent diabetes. Diabetes 1982;31:795–801.
198 Davidson MB: The effect of aging on carbohydrate metabolism: a review of the English literature and a practical approach to the diagnosis of diabetes mellitus in the elderly. Metabolism 1979;28:688–705.
199 Fink RI, Kolterman OG, Griffin J, Olefsky JM: Mechanisms of insulin resistance in aging. J Clin Invest 1983;71:1523–1535.
200 Spence JC: Blood sugar tolerance and age. Q J Med 1920–21;14:314–326.
201 Chen M, Bergman RN, Pacini G, Porte D Jr: Pathogenesis of age-related glucose intolerance in man: insulin resistance and decreased beta-cell function. J Clin Endocrinol Metab 1985;60:13–20.
202 Gelfand RA, Barrett EJ: Effect of physiologic hyperinsulinemia on skeletal muscle protein synthesis and breakdown in man. J Clin Invest 1987;80:1–6.
203 Stump CS, Short KR, Bigelow ML, Schimke JM, Nair KS: Effect of insulin on human skeletal muscle mitochondrial ATP production, protein synthesis, and mRNA transcripts. Proc Natl Acad Sci USA 2003;100:7996–8001.
204 Sambuy Y, Bittles AH: The effects of in vitro ageing on the composition of the intracellular free amino acid pool of human diploid fibroblasts. Mech Age Dev 1982;20:279–287.
205 Rudman D, Mattson DE, Feller AG, Cotter R, Johnson RC: Fasting plasma amino acids in elderly men. Am J Clin Nutr 1989;49:559–566.
206 Dardevet D, Sornet C, Bayle G, Prugnaud J, Pouyet C, Grizard J: Postprandial stimulation of muscle protein synthesis in old rats can be restored by a leucine-supplemented meal. J Nutr 2002;132:95–100.
207 Rieu I, Sornet C, Bayle G, Prugnaud J, Pouyet C, Balage M, Papet I, Grizard J, Dardevet D: Leucine-supplemented meal feeding for ten days beneficially affects postprandial muscle protein synthesis in old rats. J Nutr 2003;133:1198–1205.
208 Combaret L, Dardevet D, Rieu I, Pouch MN, Bechet D, Taillandier D, Grizard J, Attaix D: A leucine-supplemented diet restores the defective postprandial inhibition of proteasome-dependent proteolysis in aged rat skeletal muscle. J Physiol 2005;569:489–499.
209 Fiatarone MA, O'Neill EF, Ryan ND, Clements KM, Solares GR, Nelson ME, Roberts SB, Kehayias JJ, Lipsitz LA, Evans WJ: Exercise training and nutritional supplementation for physical frailty in very elderly people. N Engl J Med 1994;330:1769–1775.
210 Fujita S, Volpi E: Amino acids and muscle loss with aging. J Nutr 2006;136:277S–280S.
211 Paddon-Jones D, Wolfe RR, Ferrando AA: Amino acid supplementation for reversing bed rest and steroid myopathies. J Nutr 2005;135:1809S–1812S.
212 Dioguardi FS: Wasting and the substrate-to-energy controlled pathway: a role for insulin resistance and amino acids. Am J Cardiol 2004;93:6A–12A.
213 Young VR, Bier DM: Amino acid requirements in the adult human: how well do we know them? J Nutr 1987;117:1484–1487.
214 Mosoni L, Houlier ML, Mirand PP, Bayle G, Grizard J: Effect of amino acids alone or with insulin on muscle and liver protein synthesis in adult and old rats. Am J Physiol 1993;264:E614–E620.
215 Arnal MA, Mosoni L, Boirie Y, Houlier ML, Morin L, Verdier E, Ritz P, Antoine JM, Prugnaud J, Beaufrere B, Mirand PP: Protein pulse feeding improves protein retention in elderly women. Am J Clin Nutr 1999;69:1202–1208.
216 Lewis SE, Kelly FJ, Goldspink DF: Pre- and postnatal growth and protein turnover in smooth muscle, heart and slow- and fast-twitch skeletal muscles of the rat. Biochem J 1984;217:517–526.
217 Schoeller DA, Buchholz AC: Energetics of obesity and weight control: does diet composition matter? J Am Diet Assoc 2005;105:S24–S28.
218 Heaney RP, Layman DK: Amount and type of protein influences bone health. Am J Clin Nutr 2008;87:1567S–1570S.
219 Aquilani R, Opasich C, Gualco A, Verri M, Testa A, Pasini E, Viglio S, Iadarola P, Pastoris O, Dossena M, Boschi F: Adequate energy-protein intake is not enough to improve nutritional and metabolic status in muscle-depleted patients with chronic heart failure. Eur J Heart Fail 2008;10:1127–1135.
220 Scognamiglio R, Testa A, Aquilani R, Dioguardi FS, Pasini E: Impairment in walking capacity and myocardial function in the elderly: is there a role for nonpharmacologic therapy with nutritional amino acid supplements? Am J Cardiol 2008;101:78E–81E.
221 Aquilani R, Viglio S, Iadarola P, Opasich C, Testa A, Dioguardi FS, Pasini E: Oral amino acid supplements improve exercise capacities in elderly patients with chronic heart failure. Am J Cardiol 2008;101:104E–110E.

222 Scarabelli TM, Townsend PA, Chen Scarabelli C, Yuan Z, McCauley RB, Di Rezze J, Patel D, Putt J, Allebban Z, Abboud J, Chilukuri K, Gardin J, Saravolatz L, Knight RA, Latchman DS, Stephanou A: Amino acid supplementation differentially modulates STAT1 and STAT3 activation in the myocardium exposed to ischemia/reperfusion injury. Am J Cardiol 2008;101:63E–68E.
223 Chen Scarabelli C, McCauley RB, Yuan Z, Di Rezze J, Patel D, Putt J, Raddino R, Allebban Z, Abboud J, Scarabelli GM, Chilukuri K, Gardin J, Saravolatz L, Faggian G, Mazzucco A, Scarabelli TM: Oral administration of amino acidic supplements improves protein and energy profiles in skeletal muscle of aged rats: elongation of functional performance and acceleration of mitochondrial recovery in adenosine triphosphate after exhaustive exertion. Am J Cardiol 2008;101:42E–48E.
224 Pasini E, Scarabelli TM, D'Antona G, Dioguardi FS: Effect of amino acid mixture on the isolated ischemic heart. Am J Cardiol 2004;93:30A–34A.
225 Aquilani R, Scocchi M, Iadarola P, Franciscone P, Verri M, Boschi F, Pasini E, Viglio S: Protein supplementation may enhance the spontaneous recovery of neurological alterations in patients with ischaemic stroke. Clin Rehabil 2008;22:1042–1050.
226 Aquilani R, Scocchi M, Boschi F, Viglio S, Iadarola P, Pastoris O, Verri M: Effect of calorie-protein supplementation on the cognitive recovery of patients with subacute stroke. Nutr Neurosci 2008;11:235–240.
227 Paddon-Jones D, Short KR, Campbell WW, Volpi E, Wolfe RR: Role of dietary protein in the sarcopenia of aging. Am J Clin Nutr 2008;87:1562S–1566S.
228 Kimball SR, Jefferson LS: Regulation of protein synthesis by branched-chain amino acids. Curr Opin Clin Nutr Metab Care 2001;4:39–43.
229 Layman DK: The role of leucine in weight loss diets and glucose homeostasis. J Nutr 2003;133:261S–267S.
230 Weigle DS, Breen PA, Matthys CC, Callahan HS, Meeuws KE, Burden VR, Purnell JQ: A high-protein diet induces sustained reductions in appetite, ad libitum caloric intake, and body weight despite compensatory changes in diurnal plasma leptin and ghrelin concentrations. Am J Clin Nutr 2005;82:41–48.
231 Layman DK, Baum JI: Dietary protein impact on glycemic control during weight loss. J Nutr 2004;134:968S–973S.
232 Gannon MC, Nuttall FQ, Saeed A, Jordan K, Hoover H: An increase in dietary protein improves the blood glucose response in persons with type 2 diabetes. Am J Clin Nutr 2003;78:734–741.
233 Pellegrino MA, Brocca L, Dioguardi FS, Bottinelli R, D'Antona G: Effects of voluntary wheel running and amino acid supplementation on skeletal muscle of mice. Eur J Appl Physiol 2005;93:655–664.
234 Guarente L: Mitochondria – a nexus for aging, calorie restriction, and sirtuins? Cell 2008;132:171–176.
235 Marzetti E, Anne Lees H, Eva Wohlgemuth S, Leeuwenburgh C: Sarcopenia of aging: underlying cellular mechanisms and protection by calorie restriction. Biofactors 2009;35:28–35.
236 Taaffe DR: Sarcopenia–exercise as a treatment strategy. Aust Fam Physician 2006;35:130–134.
237 Lin J, Wu H, Tarr PT, Zhang CY, Wu Z, Boss O, Michael LF, Puigserver P, Isotani E, Olson EN, Lowell BB, Bassel-Duby R, Spiegelman BM: Transcriptional co-activator PGC-1 alpha drives the formation of slow-twitch muscle fibres. Nature 2002; 418:797–801.
238 Sandri M, Lin J, Handschin C, Yang W, Arany ZP, Lecker SH, Goldberg AL, Spiegelman BM: PGC-1 alpha protects skeletal muscle from atrophy by suppressing FoxO3 action and atrophy-specific gene transcription. Proc Natl Acad Sci USA 2006;103: 16260–16265.
239 Handschin C, Kobayashi YM, Chin S, Seale P, Campbell KP, Spiegelman BM: PGC-1alpha regulates the neuromuscular junction program and ameliorates Duchenne muscular dystrophy. Genes Dev 2007;21:770–783.
240 Anderson R, Prolla T: PGC-1alpha in aging and anti-aging interventions. Biochim Biophys Acta 2009;1790:1059–1066.
241 Wenz T, Rossi SG, Rotundo RL, Spiegelman BM, Moraes CT: Increased muscle PGC-1alpha expression protects from sarcopenia and metabolic disease during aging. Proc Natl Acad Sci USA 2009;106: 20405–20410.
242 Cunningham JT, Rodgers JT, Arlow DH, Vazquez F, Mootha VK, Puigserver P: mTOR controls mitochondrial oxidative function through a YY1-PGC-1alpha transcriptional complex. Nature 2007;450: 736–740.
243 Ramanathan A, Schreiber SL: Direct control of mitochondrial function by mTOR. Proc Natl Acad Sci USA 2009;106:22229–22232.
244 Vander Heiden MG, Chandel NS, Schumacker PT, Thompson CB: Bcl-xL prevents cell death following growth factor withdrawal by facilitating mitochondrial ATP/ADP exchange. Mol Cell 1999;3:159–167.
245 Colombini M: VDAC: the channel at the interface between mitochondria and the cytosol. Mol Cell Biochem 2004;256/257:107–115.
246 Desai BN, Myers BR, Schreiber SL: FKBP12-rapamycin-associated protein associates with mitochondria and senses osmotic stress via mitochondrial dysfunction. Proc Natl Acad Sci USA 2002;99:4319–4324.

247 Bai X, Ma D, Liu A, Shen X, Wang QJ, Liu Y, Jiang Y: Rheb activates mTOR by antagonizing its endogenous inhibitor, FKBP38. Science 2007;318:977–980.
248 Uhlenbrock K, Weiwad M, Wetzker R, Fischer G, Wittinghofer A, Rubio I: Reassessment of the role of FKBP38 in the Rheb/mTORC1 pathway. FEBS Lett 2009;583:965–970.
249 Sato T, Nakashima A, Guo L, Tamanoi F: Specific activation of mTORC1 by Rheb G-protein in vitro involves enhanced recruitment of its substrate protein. J Biol Chem 2009;284:12783–12791.
250 Corradetti MN, Inoki K, Bardeesy N, DePinho RA, Guan KL: Regulation of the TSC pathway by LKB1:evidence of a molecular link between tuberous sclerosis complex and Peutz-Jeghers syndrome. Genes Dev 2004;18:1533–1538.

Giuseppe D'Antona, MD, PhD
Department of Physiology, Human Physiology Unit and
Interuniversity Institute of Myology (IIM), University of Pavia
Via Forlanini 6, IT–27100 Pavia (Italy)
Tel. +39 0382 987252, Fax +39 0382 987664, E-Mail gdantona@unipv.it

Mobbs CV, Hof PR (eds): Body Composition and Aging.
Interdiscipl Top Gerontol. Basel, Karger, 2010, vol 37, pp 142–156

Mitochondrial Theory of Aging in Human Age-Related Sarcopenia

Gianni Parise[a,b] · Michael De Lisio[a]

Departments of [a]Kinesiology and [b]Medical Physics and Applied Radiation Sciences, McMaster University, Hamilton, Ont., Canada

Abstract

Understanding age-related sarcopenia and, more importantly, devising counterstrategies require an intimate knowledge of the underlying mechanism(s) of sarcopenia. The mitochondrial theory of aging (MTA) has been a leading theory on aging for the last decade; however, there is relatively little information from human tissue to support or rebut the involvement of the MTA in aging skeletal muscle. It is believed that mitochondria may contribute to sarcopenia in a stochastic fashion where regions of fibers containing dysfunctional mitochondria are forced to atrophy. Resistance exercise, a known hypertrophic stimulus, has been shown to improve the mitochondrial phenotype of aged skeletal muscle. Furthermore, activation of skeletal muscle stem cells by resistance exercise may attenuate sarcopenia in two ways. First by inducing nuclear addition to postmitotic fibers, and, second, by increasing the proportion of functional mitochondria donated by muscle stem cells in a process termed 'gene shifting'. In this chapter we review the evidence supporting the MTA, the potential to attenuate the MTA with a known hypertrophic stimuli and explore the role of muscle stem cells in gene shifting to determine the connection between mitochondrial dysfunction and age-related sarcopenia.

Understanding and developing counterstrategies to attenuate sarcopenia requires some level of knowledge regarding the underlying mechanisms of sarcopenia. Over the years, numerous theories on aging have been proposed with varying levels of support for each. To date, no single theory can comprehensively account for the events associated with aging but certainly some theories have gained more attention than others. The free radical theory of aging, first proposed in 1962, implicated free radical production as playing a causative role in aging by damaging important macromolecules such as DNA, ultimately leading to the cell's demise. This theory evolved to become the mitochondrial theory of aging when it was realized that mitochondria were the primary source of cellular oxidants through electron transport chain (ETC) inefficiencies. Although the theory has its limitations, it seems the mitochondrion is widely accepted to play a central role in the aging process. This is probably due to the fact that the aging phenotype is often observed in models of aging that implicate mitochondria. For example, polymerase gamma transgenic mice are completely devoid of a functional mitochondrial DNA (mtDNA) proof reading enzyme and present with an overt aging phenotype, ultimately characterized by a shortened lifespan [1]. Additionally, the MnSOD knockout mouse, devoid of the mitochondrial isoform of the antioxidant enzyme, superoxide dismutase, also presents with overt signs of premature aging [2]. Whatever the cause of aging, it seems that mitochondria are easy targets to implicate in the underlying mechanisms of aging. This perhaps is no surprise since mitochondria are the primary metabolic organelles responsible for ATP production as well as the primary site of free radical production. It is quite easy to see that a decrease in energy production, perhaps below that of the metabolic demands of the cell, or an overproduction of free radicals could lead to cell death.

How do we define the aging phenotype of skeletal muscle? Simply put, sarcopenia refers to the age-related loss of skeletal muscle mass and strength. It is now well established that appreciable losses of skeletal muscle can begin as early as the fourth decade of life and progress at a rate of 0.5–1% per year [3]. In addition to the loss of mass, there is a documented selected atrophy of type II (high force generating) muscle fibers [4], a shift in the fiber-type composition of aging muscle towards type I (low force generating) fibers [5], and a selective loss of force generating capacity in type II fibers [6]. In addition to the loss in lean mass is an increase in intramuscular fat and connective tissue. Collectively, these observations describe the events associated with age-related sarcopenia in humans. Unraveling the mysteries of aging in human skeletal muscle has only just begun, and we and others have focused our efforts on understanding age-related sarcopenia in the context of the mitochondrial theory of aging.

Although many theories of aging of have been proposed no theory of aging has been studied as extensively as the mitochondrial theory of aging (MTA). The MTA proposes a 'vicious cycle' of events that culminates in a loss of oxidant homeostasis leading towards an oxidized cellular environment. As with any biological cycle, it is difficult to tell where the cycle begins and ends but several key events have been proposed. One is that an aging cell is associated with ETC inefficiencies and an increased production of free radicals. Increased free radical production leads to mtDNA damage, resulting in the production of dysfunctional mitochondrial proteins, and consequently, impaired ETC function. The cycle continues with an even greater production of free radicals until cell death occurs. In this chapter, we will review evidence, primarily from human research, that supports an active role for mitochondrial dysfunction in age-related sarcopenia. We will also highlight how strategies to reverse/attenuate age-related sarcopenia involve adaptations in mitochondrial function, and finally we will discuss how muscle stem cell biology and mitochondrial function may collide to offer the most promising strategy yet to combat age-related sarcopenia.

Evidence of the Mitochondrial Theory of Aging in Human Skeletal Muscle

Numerous investigations have reported age-related increases in oxidative damage in skeletal muscle. In fact, increases in lipid peroxidation [7], protein carbonylation [8], damage to nuclear DNA [9] and mitochondrial DNA [10] have all been reported in aging skeletal muscle. This shift in oxidant status is thought to be a result of ETC dysfunction leading to enhanced free radical production. It is well documented that ROS are primarily produced by mitochondria at complex I [11] and complex III [12] of the ETC during oxidative phosphorylation. For a comprehensive review of the ETC and free radical production, see Poyton et al. [13]. Additionally, it is commonly observed that antioxidant enzyme activity increases in aging skeletal muscle. It is generally accepted that age-related increases in antioxidant enzymes [14–17] are a compensatory attempt to cope with an increasingly oxidative environment. Indeed, we have also reported significant increases in skeletal muscle MnSOD and catalase activity in older as compared to younger men [18]. What is particularly interesting is that despite the increase in antioxidant capacity in skeletal muscle of older adults, cellular oxidative damage persists suggesting that the age-related adaptation in antioxidant systems is not sufficient to cope with enhanced mitochondrial oxidant production. This notion is widely accepted in the literature and is supported by data from our laboratory reporting increased DNA damage, in the form of urinary 8-OHdG, and muscle protein damage, in the form of carbonylation, in older as compared to younger men [18], as well as data from others reporting greater damage to lipids [16] and proteins [8]. The importance of oxidative damage in humans was highlighted by a study demonstrating an inverse relationship between protein carbonylation and handgrip strength, suggesting that oxidative stress is inversely related to muscle strength [19]. In support of this finding, it was reported that older adults with higher circulating uric acid, known to have potent antioxidant properties, was associated with higher handgrip and knee extensor strength [20]. Collectively, these data implicate free radical production and not a loss in antioxidant capacity, as the primary mechanism responsible for the oxidative environment in aged muscle, and implicates oxidative stress as a significant determinant of strength in older adults.

The data from humans regarding ETC function during aging are somewhat difficult to interpret. The animal literature has more convincingly demonstrated an age-related decline in ETC enzyme activity [21, 22]. However, the literature on human skeletal muscle is seemingly more complicated for reasons discussed below. At this time there are at least as many reports demonstrating a decrease in ETC enzyme activity as there are those demonstrating no change as a function of aging [18, 23–29]. Why the literature is so heterogeneous can largely be explained by the heterogeneity of the populations studied. 'Aging' is a difficult term to define and as such it seems that anybody over the age 50 is sufficiently old to be considered an 'older adult'. Indeed, the age range reported in the literature is expansive and older age groups can range anywhere between 50 and 90 years. The lack of a precise definition of 'aging' is likely the primary reason as to why there is little consensus on human skeletal muscle ETC activity. In support of this hypothesis, it has been reported that motor unit number remains relatively consistent through life until the age of 60 years before significantly declining [30], and muscle fiber number remains unchanged until about 50 years but declines by 50% between the ages of 50 and 80 years [31]. This suggests that even amongst older adults (i.e. 50–90 years) there is significant variability, which is likely dependent on the degree of frailty. Personal communication with Dr. Tarnopolsky regarding a study in which his group compared young, old and frail old revealed that, in fact, mitochondrial ETC protein subunit expression was not different between young and old but was significantly different between young and frail old and old and frail old. This is striking because frailty is largely accounted for by lifestyle decisions, such as a lack of physical activity [32]. A second possibility for the heterogeneity in human literature may be attributable to the assay methodology. In most studies measuring ETC enzyme activity, a muscle biopsy is homogenized and maximal enzyme activity is measured. The problem this poses is that all fiber types and all fibers are mixed into one total homogenate, which does not account for between fiber-type differences or between fiber differences. The importance of this limitation became evident when single fiber analysis demonstrated that advanced age was associated with a greater proportion of single fibers that were completely devoid of cytochrome oxidase (COX) activity evidenced through histochemical analysis [33–35]. This suggested that it may be possible that whole muscle homogenate ETC activity may appear completely normal while indeed there may be single fibers that are completely deficient of ETC function.

This notion is consistent with observations of accumulated mtDNA deletions and mutations with aging. Mitochondrial DNA is a double stranded circular structure that is approximately 16.5 kb, and codes for 13 mitochondrial proteins, all involved in oxidative phosphorylation [36]. Each mitochondrion contains between 10 and 1,000 copies of mtDNA and is thought to be vulnerable to mutation [36]. The reasons for this vulnerability include a complete lack of introns, which means every oxidative insult affects a coding region of DNA. A second reason is a lack of protective histones, and a third reason is the proximity of mtDNA to the oxidant-producing sites in mitochondria [37]. Another primary tenet of the MTA is the accumulation of mtDNA deletions and mutations as a function of age [38]. It is thought that aberrations of mtDNA ultimately lead to dysfunctional mitochondrial proteins and, as a result, deficiencies in electron transport. Linnane et al. [10] first described the presence of mtDNA deletions in aging muscle; however, these authors reported levels of deletions that amounted to less than 0.02% of all mtDNA. This begged the question of whether the accumulation of mtDNA deletions as a function of age could actually contribute to the 'aging' phenotype or whether these deletions were simply artefact. As in the case of ETC enzyme activity the limitation of these early studies was that the measurement of mtDNA deletions was performed in whole muscle homogenates. Interestingly, around the turn of the decade a series of studies by

Aiken and colleagues highlighted the importance of examining mtDNA deletions in single fibers. These studies were performed in rats, monkeys, and humans all demonstrating strikingly similar results across the age span [33–35]. The approach in these studies was to stain muscle cross sections for COX and succinate dehydrogenase (SDH) activity. The relevance of these two specific enzymes is that COX contains mitochondrial encoded subunits, while subunits of SDH are encoded entirely by nuclear DNA. Sections were screened for fibers containing COX– and SDH++ fibers. Collectively, this particular profile of staining was indicative of a fiber that had lost ETC activity due to mtDNA dysfunction, while the nucleus attempted to restore the normal energy status of the cell by upregulating nuclear encoded subunits of the ETC. Using laser capture microdissection these fibers were analyzed for mtDNA deletions and it was discovered that COX–/SDH++ fibers were nearly entirely deficient of normal mtDNA. This suggested that although at the whole tissue level, mtDNA mutations only accumulated to a minute proportion of the total mtDNA, when analyzed at the single fiber level the majority of mtDNA could be deleted in regions of ETC dysfunction. It has been argued that one could not conclude that the mtDNA deletion led to the ETC dysfunction because it is possible that the presence of mtDNA deletions were merely the consequence of a cell undergoing death. Although this is plausible, the fact that the nucleus significantly upregulated subunits of the ETC in the presence of widespread mtDNA deletion would argue that indeed mtDNA deletions were causative in age-related ETC dysfunction. Although preliminary, recent data suggests that large scale mtDNA deletions (i.e. 4,977 bp 'common deletion') may not just be an indicator of mtDNA damage with age but may play an active role in interfering with oxidative phosphorylation and inducing apoptosis [39]. This hypothesis requires further investigation. The primary question however, and indeed the focus of this chapter, is how could mitochondrial dysfunction lead to age-related sarcopenia?

It is worth stepping back and considering how age-related sarcopenia may manifest. The simplest theory is that over the course of time protein breakdown exceeds that of protein synthesis leading to a net negative protein balance ultimately causing the loss of skeletal muscle proteins and hence muscle mass. This has indeed been shown to occur [40–43]. However, alternative hypotheses, not mutually exclusive from the argument of declining protein synthesis with age, have been proposed [35, 44]. Perhaps the most intriguing finding reported was the fact that in regions demonstrating COX–/SDH++ fibers and containing a complete loss of normal mtDNA the fiber was significantly atrophied relative to adjacent fibers expressing normal ETC activity. Upon further investigation it was revealed that even within a single fiber the ETC deficient regions were confined to only a proportion of that fiber. In other words, within the same fiber, there were regions of ETC deficiency and significant mtDNA deletion associated with fiber atrophy. However, directly adjacent to these regions, within the same fiber, were regions that were completely normal for ETC function, mtDNA, and fiber size. These observations suggest that perhaps age-

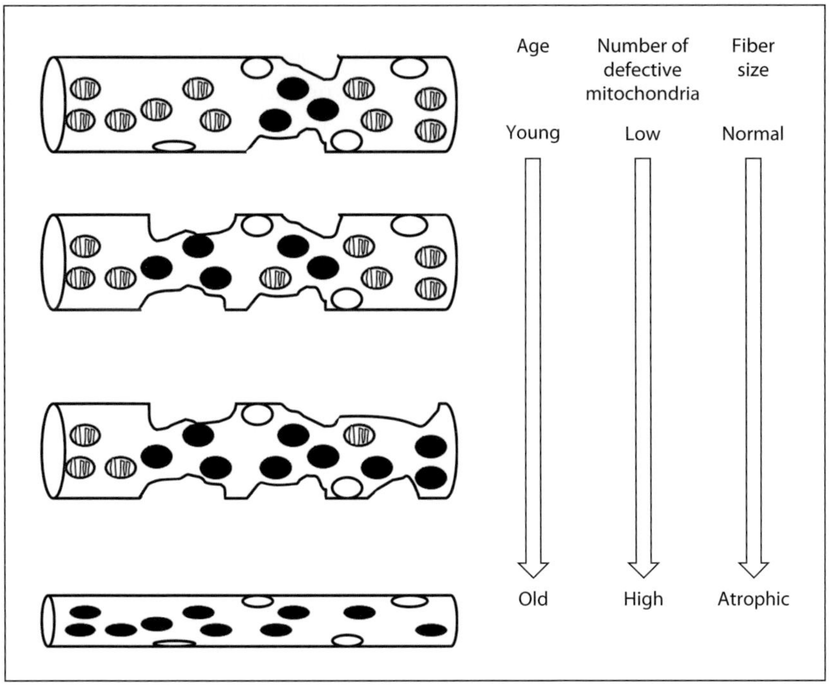

Fig. 1. Model of muscle atrophy associated with dysfunctional mitochondria. The figure outlines a model of how mitochondria may contribute to sarcopenia. Over the course of a lifetime, regions of fibers may accumulate dysfunctional mitochondria, leading to muscle fiber atrophy. Open circles represent nuclei, black circles represent dysfunctional mitochondria, open circles with black lines represent functional mitochondria.

related sarcopenia is much more complex than first thought. It is possible that over the course of a lifetime, these 'hotspots' within a fiber flare up and atrophy eventually leading to the death of a region of a fiber, suggesting a stochastic model of skeletal muscle atrophy. Over the course of decades the loss of these regions may lead to age-related muscle atrophy, while never demonstrating a significant loss of ETC function or a significant accumulation of mtDNA deletions at the whole muscle level (fig. 1). If indeed, mitochondrial dysfunction could lead to skeletal muscle atrophy then does resistance exercise (RE) training, a known hypertrophic stimulus, lead to beneficial mitochondrial adaptations? This is a question that will be considered in the following section.

Resistance Exercise and Mitochondrial Adaptation in Human Skeletal Muscle

Traditionally, the induction of mitochondrial adaptation through exercise was thought to be achieved solely through endurance-type exercise. More recently, however, convincing evidence suggests that anaerobic-type activity and even RE may be a potent stimulus for mitochondrial adaptation [45]. Perhaps the simplest clue suggesting that mitochondria adapt to RE is the fact that one would expect a decrease in skeletal muscle mitochondrial density if indeed mitochondria did not at all adapt to RE. In fact, in aging human muscle, the complete opposite has been observed. Six months of RE training in older adults (~70 years) has been shown to result in an approximate 30% increase in mitochondrial volume density [46]. Irrespective of the controversy underlying exercise-specific adaptation, what is indisputable is the fact that regular RE in older adults is an effective strategy to increase strength [47, 48], significantly impact body composition by inducing muscle fiber hypertrophy [49, 50], and data from our lab even suggest that regular RE can increase oxidative capacity [51]. This final piece of evidence came from experiments in our lab designed to examine whether, indeed, RE training could impact the underlying mechanism(s) of aging. The assumption of course is that the MTA is, in fact, part of the underlying cause of skeletal muscle aging.

Before discussing the impact of RE on aspects of the MTA it is worth stating that, at least in the opinion of this author, RE training is the single most effective strategy for attenuating age-related sarcopenia. Many investigations have demonstrated that older adults and even frail elderly individuals can increase strength and muscle hypertrophy by engaging in a regular RE program [42, 47, 49, 50, 52–58]. Perhaps more importantly is the improvement in functionality associated with RE [47, 48, 59] and the reduction in morbidity [60] and mortality [61, 62]. Given such striking benefits, our interest was to determine whether specific aspects of the MTA were influenced by RE or whether the adaptations associated with RE were simply compensatory with no affect on the underlying mechanism(s) of aging. Surprisingly, there is very little information on oxidant status in skeletal muscle in elderly individuals following RE. Vincent et al. [63] reported that RE training in elderly individuals led to reduced lipid peroxidation and decreased levels of hydroperoxide following an aerobic exercise challenge in trained older adults. We reported that whole body RE led to a reduction in urinary 8-OHdG excretion, a marker of oxidative DNA damage, suggesting that RE led to a shift in oxidant status [51]. Surprisingly, this was not accompanied by an increase in antioxidant enzyme protein expression. However, it is has been suggested that antioxidant upregulation in response to exercise may be achieved through post-translational modifications of antioxidant proteins [64]. Data from trained rats demonstrated that skeletal muscle CuZnSOD activity was significantly upregulated over sedentary control rats, however CuZnSOD protein levels remained unchanged [64]. Indeed, upon further investigation we reported a 75% increase in CuZnSOD activity and an 82% increase in catalase activity following training in older adults [65]. Interestingly, we did not observe a change in MnSOD activity suggesting that the adaptive response was compartmentalized. CuZnSOD and catalase are both cytosolic enzymes while MnSOD is compartmentalized to the mitochondria. Since changes were only observed in the enzymes compartmentalized to the cytosol it is fair to assume that the majority of the oxidative stress associated with the RE occurred in the cytosol.

This notion is supported by several lines of evidence. First, it is known that resistance type exercise is associated with an inflammatory response that peaks 24–48 h following an exercise bout [66, 67]. This inflammatory response is associated with the oxidant producing enzyme activity of NADPH oxidase [68]. A second possible reason for the compartmentalized antioxidant response is the conversion of xanthine dehydrogenase to xanthine oxidase, which occurs with the loss of calcium homeostasis and results in the production of potent oxidants [69, 70]. Resistance type exercise is often associated with membrane damage and loss of calcium homeostasis, which could ultimately promote oxidant production. These observations would serve to shift oxidant status to a more oxidized cellular environment and actually promote the progression of the MTA. Since mitochondrial function and ETC enzyme activity are central to the MTA then it would be assumed that mitochondrial function would improve with RE training. We did not observe any changes in complex I+III, complex II+III, or citrate synthase following the training period; however, we did observe a significant upregulation in COX activity, and a significant increase in the COX complex I+III ratio [51]. This is of particular importance since it has been suggested that complex IV is an important enzyme in maintaining electron flow [71] and that COX activity is inversely related to electron leakage and oxidant production [72]. Collectively, our data suggest that RE may reduce oxidant stress through multiple mechanisms, which may directly interfere with the propagation of the MTA.

Muscle Stem Cells and Mitochondria: A Crossroads in the Treatment of Age-Related Sarcopenia?

One aspect of the MTA that remains relatively unexplored is whether RE can influence the accumulation of mtDNA deletions and mutations. With short-term training, it appears that the profile of normal and deleted mtDNA cannot be altered. Over the course of 14 weeks, we did not observe any changes in the content of normal mtDNA or deleted mtDNA in whole muscle homogenates [51]. However, it is likely that 14 weeks is not a sufficient period of intervention to make such observations. In populations of patients with mitochondrial disease, however, it has been rather convincingly demonstrated that exercise-training can indeed shift the profile of mtDNA towards normal [73]. This concept was referred to as 'gene shifting', and since aging has been likened to a mild form of mitochondrial disease, older adults may stand to gain from this phenomenon. The potential for gene shifting is dependent on two main assumptions. The first is that muscle stem cell mtDNA acquires deletions and mutations at a much slower rate than mtDNA in post mitotic fibers. This has been attributed to the much lower metabolic rate of a quiescent muscle stem cell, as compared to that of a postmitotic muscle fiber. There are many examples of a link between metabolic rate and oxidant production, although there are exceptions to this rule [74]. Thus, it is conceivable that mtDNA from muscle stem cells is relatively protected over the lifespan as compared to mtDNA from muscle fibers. The intriguing aspect of this theory is that it represents the collision of two worlds at the center of age-related sarcopenia. On the one hand, this theory depends on a functional muscle stem cell population, which has been questioned as a potential mechanism underlying age-related sarcopenia. On the other hand, this theory is dependent on muscle stem cell mitochondria being devoid of mtDNA deletions and mutations, relative to mitochondria in the postmitotic muscle fiber. To date, this has not been established in the published literature in human skeletal muscle. To this end, however, the research group of Dr. Mark Tarnopolsky has been testing the concept of gene shifting in skeletal muscle of older adults. Through personal communication, he has revealed that indeed when muscle stem cells are cultured from biopsies of older adults, and screened for mtDNA deletions, there is a complete absence of mtDNA deletions as compared to muscle fibers from the same subjects. This manuscript is currently in preparation and represents the proof of principle that gene shifting may be a plausible strategy for the treatment of age-related sarcopenia.

The general concept of the gene-shifting hypothesis is that when proliferating muscle stem cells are induced to differentiate they donate all subcellular organelles, including mitochondria, to the muscle fiber milieu [73]. If, indeed, muscle stem cell mitochondria contain fewer mtDNA deletions then this will shift the muscle fiber mtDNA profile towards normal. In conditions such as mitochondrial disease where mutated and deleted mtDNA are much higher than in aging muscle the benefits of exercise training are realized in a much shorter period of time. However, in all likelihood, long-term training studies will be necessary to verify the potential of 'gene shifting' for aging populations.

Ultimately, the gene shifting paradigm is not plausible without the appropriate function of muscle stem cells. Muscle stem cells have become a target in the quest to understand age-related sarcopenia. Literature from humans on aging muscle stem cells has focused on enumeration of the muscle stem cell pool, and at the moment, the literature is equivocal. There are nearly as many reports that muscle stem cell number is reduced with age [75–78] as there are those that suggest the muscle stem cell pool does not change as a function of age [77, 79–83]. Another theory suggests that the size of the muscle stem cell pool does not change with age, but rather in advanced age the ability to activate the stem cell pool becomes impaired [84]. A limited pool of data from humans suggests that human satellite cells may be impaired in their ability to respond to a hypertrophic stimulus, demonstrating attenuated increases in satellite cells following exercise [79–80]. Additionally, some data suggests that satellite cells from older adults may be functionally impaired [85–86]. Thus, to make 'gene shifting' an effective strategy we must identify how to maximally activate the muscle stem cell pool.

Muscle Stem Cells

Muscle stem cells, also known as satellite cells, were first described in 1961 [87]. These cells were named 'satellite cells' due to their position on the muscle fiber between the sarcolemma and the basal lamina. Although these cells represent only 2% of all myonuclei they have the remarkable ability to completely regenerate a muscle within weeks of degeneration [88]. In addition, skeletal muscle growth is severely impaired without the presence of a functional satellite cell population [89]. Since their description over half a century ago, these cells remain somewhat of an enigma. The progression of these cells from the quiescent state through activation, proliferation and terminal differentiation is referred to as the myogenic program and is regulated by a group of transcription factors known as the myogenic regulatory factors (MRF) [90]. The MRFs are a group of transcription factors containing a basic-helix-loop-helix domain (bHLH) of which Myf5 and MyoD are known to control proliferation while Myogenin and Mrf4 are thought to direct differentiation. Most of the characterization of this work has been done in animals, but work in humans loosely supports the animal data. We have recently determined the MRF expression profile over a 4-day period following a single bout of damaging eccentric contractions of the vastus lateralis muscle. Our data demonstrate a peak in MyoD and Myf5 gene expression 4 and 24 h, respectively, following exercise, while Mrf4 and Myogenin peaked at 72 h following damage supporting the notion that Myf5 and MyoD are important in the early events (i.e. proliferation) following damage while Mrf4 and myogenin are important in the later events following damage (i.e. differentiation) [91]. The time course for MRF expression in older adults following damage remains unknown.

Although the transcriptional networks orchestrating the progression of satellite cells through the myogeninc program are relatively well characterized, the cues that induce the MRFs are poorly understood. In humans, we know little regarding the regulation of satellite cells and hence whether aging has an impact on the factors that control the MRFs is completely unknown. Interestingly, spliced variants of insulin-like growth factor (IGF) have been implicated in growth [92]. We have recently shown that IGF protein colocalizes to satellite cells following damage and that IGF-1Ea and IGF-1Eb gene expression correlate well to myogenin and Mrf4 expression suggesting a temporal relationship between IGF-1Ea, IGF-1Eb and satellite cell differentiation. Additionally, we reported that a third spliced variant of IGF-1, mechano-growth factor, gene expression correlated well with Myf5 expression suggesting a temporal relationship between mechano-growth factor and satellite cell proliferation [91]. More recently, we have reported an important role for IL-6 in the regulation of human muscle satellite cells. Following a single bout of damaging eccentric contractions we report that IL-6 protein colocalized to satellite cells and although negligibly present at rest was expressed in approximately 80% of all satellite cells 24 h following damage. Interestingly, IL-6 colocalization was almost completely abolished by 72 h suggesting that IL-6 may be temporally involved in the proliferative phase of the myogenic program [93]. The mysteries

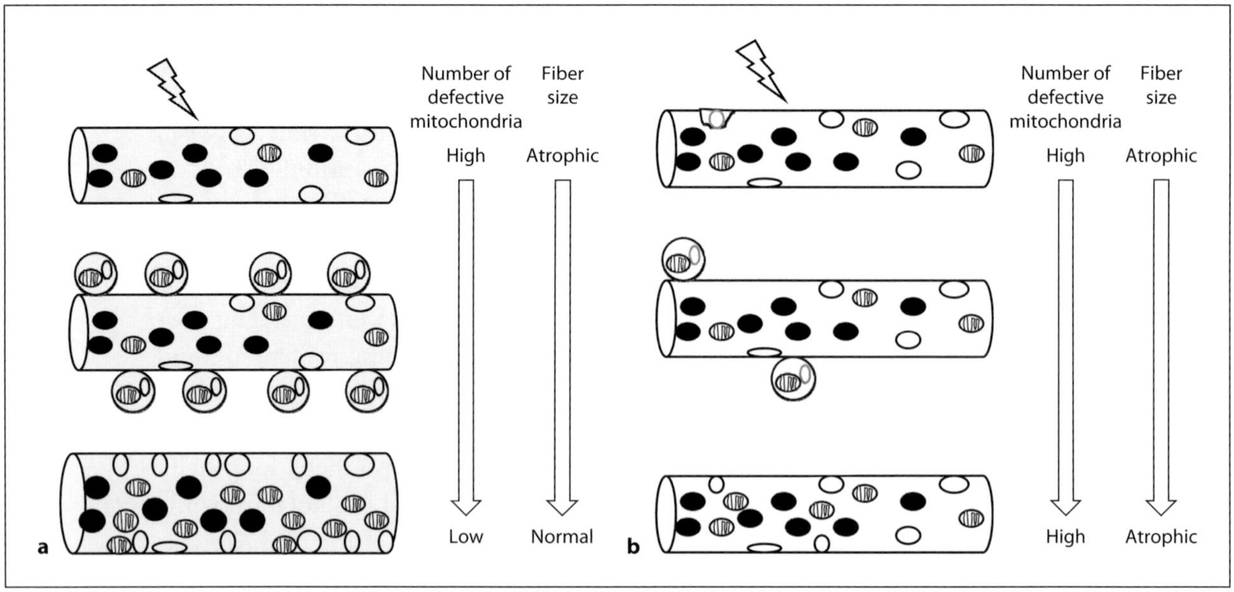

Fig. 2. a Gene shifting paradigm with functional satellite cells. Three fibers are presented with nuclei as open circles, functional mitochondria are open with black lines, dysfunctional mitochondria are black, and satellite cells are found outside the fiber and contain a mitochondrion and a nuclei. Following an insult, satellite cells are activated, proliferate and donate their nuceli and miotochondria to the fibers. The result is increased fiber growth, and a decrease in the proportion of dysfunctional mitochondria. **b** Gene shifting paradigm with dysfunctional satellite cells. This figure depicts the gene shifting paradigm in the absence of a functional satellite cell population. Following an insult, the satellite cell population cannot respond and the result is no change in the mitochondrial profile of the fiber and no associated muscle growth.

of human satellite cell regulation are slowly being revealed and it is imperative that these factors are examined in aging skeletal muscle to identify deficits in the regulation of satellite cells from older adults. Although the concept of 'gene shifting' brings great promise, without the effective function of muscle stem cells gene shifting may not be a reality (fig. 2). It is a provocative theory that mitochondrial function could determine the onset and progression of age-related sarcopenia, and although not entirely intuitive, perhaps the appropriate function of muscle stem cells is not only necessary for nuclear support of growth and repair but equally important for maintaining mitochondrial function, which may indirectly impact muscle size.

Perspectives

Age-related sarcopenia is a significant problem that has a major impact on the quality of life of our aging population and an enormous impact on the healthcare system.

Devising strategies to counter the effects of age-related sarcopenia is a significant priority, but requires an understanding of the mechanism(s) that underlie the pathophysiology of sarcopenia. The MTA has been recognized, and is widely accepted, as a viable theory on the aging process. We and others have shown that indeed aspects of the MTA are manifested in human skeletal muscle, and work in single fibers has drawn a convincing argument that mitochondrial dysfunction is associated with a loss of muscle fiber cross-sectional area. Perhaps more important is the fact that components of the MTA can be significantly attenuated by known hypertrophic stimuli such as RE. In attempting to tackle the problem of age-related sarcopenia it is equally important to realize that sarcopenia is likely a multifactorial process and much more complicated than we know. Nonetheless, it appears that mitochondria play a central role in skeletal muscle dysfunction and atrophy, and the road to rejuvenation may involve the recovery and maintenance of mitochondrial function.

References

1 Copeland WC: The mitochondrial DNA polymerase in health and disease. Subcell Biochem 2010;50:211–222.

2 Jang YC, Remmen VH: The mitochondrial theory of aging: insight from transgenic and knockout mouse models. Exp Gerontol 2009;44:256–260.

3 Janssen I, et al: Skeletal muscle mass and distribution in 468 men and women aged 18–88 yr. J Appl Physiol 2000;89:81–88.

4 Lexell J: Human aging, muscle mass, and fiber type composition. J Gerontol A Biol Sci Med Sci 1995;50: 11–16.

5 Korhonen MT, et al: Aging, muscle fiber type, and contractile function in sprint-trained athletes. J Appl Physiol 2006;101:906–917.

6 Trappe S, et al: Single muscle fiber contractile properties in young and old men and women. J Physiol 2003;552:47–58.

7 Pansarasa O, et al: 'Oxidative stress': effects of mild endurance training and testosterone treatment on rat gastrocnemius muscle. Eur J Appl Physiol 2002; 87:550–555.

8 Rousset S, et al: The biology of mitochondrial uncoupling proteins. Diabetes 2004;53(suppl 1): S130–S135.

9 Fano G, et al: Age and sex influence on oxidative damage and functional status in human skeletal muscle. J Muscle Res Cell Motil 2001;22:345–351.

10 Linnane AW, et al: Mitochondrial DNA mutations as an important contributor to ageing and degenerative diseases. Lancet 1989;i:642–645.

11 Herrero A, Barja G: H2O2 production of heart mitochondria and aging rate are slower in canaries and parakeets than in mice: sites of free radical generation and mechanisms involved. Mech Ageing Dev 1998;103:133–146.

12 Muller FL, Liu Y, Van Remmen H: Complex III releases superoxide to both sides of the inner mitochondrial membrane. J Biol Chem 2004;279:49064–49073.

13 Poyton RO, Ball KA, Castello PR: Mitochondrial generation of free radicals and hypoxic signaling. Trends Endocrinol Metab 2009;20:332–340.

14 Leeuwenburgh C, et al: Aging and exercise training in skeletal muscle: responses of glutathione and antioxidant enzyme systems. Am J Physiol 1994; 267(2 Pt 2):R439–R445.

15 Luhtala TA, et al: Dietary restriction attenuates age-related increases in rat skeletal muscle antioxidant enzyme activities. J Gerontol 1994;49:B231–B238.

16 Pansarasa O, et al: Age and sex differences in human skeletal muscle: role of reactive oxygen species. Free Radic Res 2000;33:287–293.

17 Marzani B, et al: Human muscle aging: ROS-mediated alterations in rectus abdominis and vastus lateralis muscles. Exp Gerontol 2005;40:959–965.

18 Gianni P, et al: Oxidative stress and the mitochondrial theory of aging in human skeletal muscle. Exp Gerontol 2004;39:1391–1400.

19 Howard C, et al: Oxidative protein damage is associated with poor grip strength among older women living in the community. J Appl Physiol 2007;103:17–20.

20 Macchi C, et al: Higher circulating levels of uric acid are prospectively associated with better muscle function in older persons. Mech Ageing Dev 2008; 129:522–527.
21 Hepple RT, et al: Caloric restriction protects mitochondrial function with aging in skeletal and cardiac muscles. Rejuvenation Res 2006;9:219–222.
22 Hagen JL, et al: Skeletal muscle aging in F344BN F1-hybrid rats. I. Mitochondrial dysfunction contributes to the age-associated reduction in VO2max. J Gerontol A Biol Sci Med Sci 2004;59:1099–1110.
23 Barrientos A, et al: Absence of relationship between the level of electron transport chain activities and aging in human skeletal muscle. Biochem Biophys Res Commun 1996;229:536–539.
24 Boffoli D, et al: Decline with age of the respiratory chain activity in human skeletal muscle. Biochim Biophys Acta 1994;1226:73–82.
25 Chretien D, et al: Biochemical parameters for the diagnosis of mitochondrial respiratory chain deficiency in humans, and their lack of age-related changes. Biochem J 1998;329:249–254.
26 Cooper JM, Mann VM, Schapira AH: Analyses of mitochondrial respiratory chain function and mitochondrial DNA deletion in human skeletal muscle: effect of ageing. J Neurol Sci 1992;113:91–98.
27 Hsieh RH, et al: Age-dependent respiratory function decline and DNA deletions in human muscle mitochondria. Biochem Mol Biol Int 1994;32:1009–1022.
28 Rasmussen UF, et al: Experimental evidence against the mitochondrial theory of aging. A study of isolated human skeletal muscle mitochondria. Exp Gerontol 2003;38:877–886.
29 Trounce I, Byrne E, Marzuki S: Decline in skeletal muscle mitochondrial respiratory chain function: possible factor in ageing. Lancet 1989;i:637–639.
30 Campbell MJ, McComas AJ, Petito F: Physiological changes in ageing muscles. J Neurol Neurosurg Psychiatry 1973;36:174–182.
31 Lexell J, Taylor CC, Sjostrom M: What is the cause of the ageing atrophy? Total number, size and proportion of different fiber types studied in whole vastus lateralis muscle from 15- to 83-year-old men. J Neurol Sci 1988;84:275–294.
32 Bortz WM 2nd: A conceptual framework of frailty: a review. J Gerontol A Biol Sci Med Sci 2002;57: M283–M288.
33 Cao Z, et al: Mitochondrial DNA deletion mutations are concomitant with ragged red regions of individual, aged muscle fibers: analysis by laser-capture microdissection. Nucleic Acids Res 2001;29: 4502–4508.
34 Wanagat J, et al: Mitochondrial DNA deletion mutations colocalize with segmental electron transport system abnormalities, muscle fiber atrophy, fiber splitting, and oxidative damage in sarcopenia. FASEB J 2001;15:322–332.
35 Bua EA, et al: Mitochondrial abnormalities are more frequent in muscles undergoing sarcopenia. J Appl Physiol 2002;92:2617–2624.
36 Wallace DC: Mitochondrial diseases in man and mouse. Science 1999;283:1482–1488.
37 Zastawny TH, et al: Comparison of oxidative base damage in mitochondrial and nuclear DNA. Free Radic Biol Med 1998;24:722–725.
38 Kopsidas G, et al: Tissue mitochondrial DNA changes: a stochastic system. Ann NY Acad Sci 2000;908:226–243.
39 Meissner C, et al: The 4977 bp deletion of mitochondrial DNA in human skeletal muscle, heart and different areas of the brain: a useful biomarker or more? Exp Gerontol 2008;43:645–652.
40 Welle S, et al: Myofibrillar protein synthesis in young and old men. Am J Physiol 1993;264(5 Pt 1): E693–E698.
41 Balagopal P, et al: Effects of aging on in vivo synthesis of skeletal muscle myosin heavy-chain and sarcoplasmic protein in humans. Am J Physiol 1997; 273:E790–E800.
42 Yarasheski KE, et al: Resistance exercise training increases mixed muscle protein synthesis rate in frail women and men >/ = 76 yr old. Am J Physiol 1999;277(1 Pt 1):E118–E125.
43 Yarasheski KE, Zachwieja JJ, Bier DM: Acute effects of resistance exercise on muscle protein synthesis rate in young and elderly men and women. Am J Physiol 1993;265:E210–E214.
44 Kopsidas G, et al: Stochastic mitochondrial DNA changes: bioenergy decline in type I skeletal muscle fibers correlates with a decline in the amount of amplifiable full-length mtDNA. Biogerontology 2002;3:29–36.
45 Gibala M: Molecular responses to high-intensity interval exercise. Appl Physiol Nutr Metab 2009;34: 428–432.
46 Jubrias SA, et al: Large energetic adaptations of elderly muscle to resistance and endurance training. J Appl Physiol 2001;90:1663–1670.
47 Fiatarone MA, et al: Exercise training and nutritional supplementation for physical frailty in very elderly people. N Engl J Med 1994;330:1769–1775.
48 Schlicht J, Camaione DN, Owen SV: Effect of intense strength training on standing balance, walking speed, and sit-to-stand performance in older adults. J Gerontol A Biol Sci Med Sci 2001;56:M281–M286.

49 Charette SL, et al: Muscle hypertrophy response to resistance training in older women. J Appl Physiol 1991;70:1912–1916.
50 Frontera WR, et al: Strength conditioning in older men: skeletal muscle hypertrophy and improved function. J Appl Physiol 1988;64:1038–1044.
51 Parise G, Brose AN, Tarnopolsky MA: Resistance exercise training decreases oxidative damage to DNA and increases cytochrome oxidase activity in older adults. Exp Gerontol 2005;40:173–180.
52 Binder EF, et al: Effects of progressive resistance training on body composition in frail older adults: results of a randomized, controlled trial. J Gerontol A Biol Sci Med Sci 2005;60:1425–1431.
53 Brose A, Parise G, Tarnopolsky MA: Creatine supplementation enhances isometric strength and body composition improvements following strength exercise training in older adults. J Gerontol A Biol Sci Med Sci 2003;58:11–19.
54 Hakkinen K, et al: Changes in agonist-antagonist EMG, muscle CSA, and force during strength training in middle-aged and older people. J Appl Physiol 1998;84:1341–1349.
55 Hakkinen K, et al: Changes in electromyographic activity, muscle fiber and force production characteristics during heavy resistance/power strength training in middle-aged and older men and women. Acta Physiol Scand 2001;171:51–62.
56 Ivey FM, et al: Effects of strength training and detraining on muscle quality: age and gender comparisons. J Gerontol A Biol Sci Med Sci 2000;55: B152–B157; discussion B158–B159.
57 Lemmer JT, et al: Age and gender responses to strength training and detraining. Med Sci Sports Exerc 2000;32:1505–1512.
58 Welle S, Totterman S, Thornton C: Effect of age on muscle hypertrophy induced by resistance training. J Gerontol A Biol Sci Med Sci 1996;51:M270–M275.
59 Hunter GR, McCarthy JP, Bamman MM: Effects of resistance training on older adults. Sports Med 2004;34:329–348.
60 Fries JF: Physical activity, the compression of morbidity, and the health of the elderly. JR Soc Med 1996;89:64–68.
61 Paffenbarger RS Jr, et al: The association of changes in physical-activity level and other lifestyle characteristics with mortality among men. N Engl J Med 1993;328:538–545.
62 Sandvik L, et al: Physical fitness as a predictor of mortality among healthy, middle-aged Norwegian men. N Engl J Med 1993;328:533–537.
63 Vincent KR, et al: Resistance exercise training attenuates exercise-induced lipid peroxidation in the elderly. Eur J Appl Physiol 2002;87:416–423.
64 Oh-ishi S, et al: Effects of endurance training on superoxide dismutase activity, content and mRNA expression in rat muscle. Clin Exp Pharmacol Physiol 1997;24:326–332.
65 Parise G, et al: Antioxidant enzyme activity is up-regulated after unilateral resistance exercise training in older adults. Free Radic Biol Med 2005;39:289–295.
66 Beaton LJ, Tarnopolsky MA, Phillips SM: Variability in estimating eccentric contraction-induced muscle damage and inflammation in humans. Can J Appl Physiol 2002;27:516–526.
67 Stupka N, et al: Gender differences in muscle inflammation after eccentric exercise. J Appl Physiol 2000;89:2325–2332.
68 Babior BM: Oxygen-dependent microbial killing by phagocytes (first of two parts). N Engl J Med 1978; 298:659–668.
69 McCord JM, Roy RS, Schaffer SW: Free radicals and myocardial ischemia. The role of xanthine oxidase. Adv Myocardiol 1985;5:183–189.
70 Kondo H, et al: Mechanism of oxidative stress in skeletal muscle atrophied by immobilization. Am J Physiol 1993;265(6 Pt 1):E839–E844.
71 Martinez M, et al: Age-related increase in oxidized proteins in mouse synaptic mitochondria. Brain Res 1996;731:246–248.
72 Kwong LK, Sohal RS: Age-related changes in activities of mitochondrial electron transport complexes in various tissues of the mouse. Arch Biochem Biophys 2000;373:16–22.
73 Taivassalo T, et al: Gene shifting: a novel therapy for mitochondrial myopathy. Hum Mol Genet 1999;8: 1047–1052.
74 Barja G: Mitochondrial oxygen consumption and reactive oxygen species production are independently modulated: implications for aging studies. Rejuvenation Res 2007;10:215–224.
75 Snow MH: The effects of aging on satellite cells in skeletal muscles of mice and rats. Cell Tissue Res 1977;185:399–408.
76 Kadi F, et al: Satellite cells and myonuclei in young and elderly women and men. Muscle Nerve 2004;29: 120–127.
77 Renault V, et al: Skeletal muscle regeneration and the mitotic clock. Exp Gerontol 2000;35:711–719.
78 Sajko S, et al: Frequency of M-cadherin-stained satellite cells declines in human muscles during aging. J Histochem Cytochem 2004;52:179–185.
79 Dreyer HC, et al: Satellite cell numbers in young and older men 24 hours after eccentric exercise. Muscle Nerve 2006;33:242–253.
80 Petrella JK, et al: Efficacy of myonuclear addition may explain differential myofiber growth among resistance-trained young and older men and women. Am J Physiol Endocrinol Metab 2006;291: E937–E946.

81 Verney J, et al: Effects of combined lower body endurance and upper body resistance training on the satellite cell pool in elderly subjects. Muscle Nerve 2008;38:1147–1154.
82 Roth SM, et al: Skeletal muscle satellite cell populations in healthy young and older men and women. Anat Rec 2000;260:351–358.
83 Hikida RS, et al: Effects of high-intensity resistance training on untrained older men. II. Muscle fiber characteristics and nucleo-cytoplasmic relationships. J Gerontol A Biol Sci Med Sci 2000;55:B347–B354.
84 Conboy IM, Rando TA: Aging, stem cells and tissue regeneration: lessons from muscle. Cell Cycle 2005;4:407–410.
85 Beccafico S, et al: Age-dependent effects on functional aspects in human satellite cells. Ann NY Acad Sci 2007;1100:345–352.
86 Lorenzon P, et al: Ageing affects the differentiation potential of human myoblasts. Exp Gerontol 2004;39:1545–1554.
87 Mauro A: Satellite cell of skeletal muscle fibers. J Biophys Biochem Cytol 1961;9:493–495.
88 Parise G, McKinnell IW, Rudnicki MA: Muscle satellite cell and atypical myogenic progenitor response following exercise. Muscle Nerve 2008;37:611–619.
89 Adams GR, et al: Cellular and molecular responses to increased skeletal muscle loading after irradiation. Am J Physiol Cell Physiol 2002;283:C1182–C1195.
90 Parise G, O'Reilly CE, Rudnicki MA: Molecular regulation of myogenic progenitor populations. Appl Physiol Nutr Metab 2006;31:773–781.
91 McKay BR, et al: Co-expression of IGF-1 family members with myogenic regulatory factors following acute damaging muscle-lengthening contractions in humans. J Physiol 2008;586:5549–5560.
92 Clemmons DR: Role of IGF-I in skeletal muscle mass maintenance. Trends Endocrinol Metab 2009;20:349–356.
93 McKay BR, et al: Association of interleukin-6 signalling with the muscle stem cell response following muscle-lengthening contractions in humans. PLoS One 2009;4:e6027.

Gianni Parise, PhD
Department of Kinesiology, McMaster University
1280 Main Street West
Hamilton, ON L8S 4K1 (Canada)
Tel. +1 905 525 9140, Fax +1 905 523 6011, E-Mail pariseg@mcmaster.ca

Mobbs CV, Hof PR (eds): Body Composition and Aging.
Interdiscipl Top Gerontol. Basel, Karger, 2010, vol 37, pp 157–174

Exercise as a Calorie Restriction Mimetic: Implications for Improving Healthy Aging and Longevity

Derek M. Huffman

Department of Medicine and Institute for Aging Research, Albert Einstein College of Medicine, Bronx, N.Y., USA

Abstract

Calorie restriction (CR) is the only paradigm that has consistently increased lifespan in a wide variety of model organisms. Many hypotheses have been proposed as the underlying mechanism, including a reduction in body size and adiposity, which is commonly observed in calorie-restricted animals. This has led to investigations as to whether similar changes in body composition produced by increasing energy expenditure via exercise can replace or enhance the benefits of reducing energy intake. The goal of this chapter is to review and discuss the evidence regarding exercise as a CR mimetic for healthy aging and longevity. In rodents, the data clearly show that exercise, regardless of body weight changes, can improve health and survival, but unlike CR, fails to extend lifespan. In humans, short-term weight loss studies show that exercise and CR produce similar improvements in disease risk factors and biomarkers of aging, while some parameters clearly benefit more with exercise. Epidemiologic evidence in humans supports exercise as a strategy to reduce the risk of morbidity and mortality, but not to extend lifespan. It is unknown whether CR can extend human lifespan, but the metabolic profile of humans engaged in long-term CR shares many similarities with calorie-restricted rodents and nonhuman primates. In conclusion, like CR, exercise can limit weight gain and adiposity, but only CR can extend lifespan. Therefore, in rodents, the ability of CR to slow aging is apparently more dependent on decreasing nutrient flux, rather than changes in energy balance and body composition.

Calorie restriction (CR) can robustly delay or prevent the onset of many age-related diseases and extend lifespan [1]. This finding has since been extended to diverse organisms including the yeast, worm, water flea, fruitfly, dog and cow [1–5]. In the laboratory, CR is generally implemented by reducing food intake 20–40% of ad libitum-fed controls [6] and has served as a useful model for discovering the potential mechanisms involved in slowing the aging process. Several hypotheses have been proposed in mediating the effects of CR on lifespan and have been reviewed in detail elsewhere [7]. One hypothesis involves the reduction in body size and adiposity that is commonly observed in mammals when caloric intake is reduced by 20–40% [8]. Indeed, throughout nature, a smaller body size within species is associated with increased lifespan; small dogs often live longer than large dogs [9], ponies live longer than thoroughbreds, and a variety of dwarf mice live longer than wild-type mice [10]. Likewise, CR prevents the accumulation of adipose tissue, and more importantly, the age-related accrual of visceral fat [11], which has been implicated as an important modulator for disease and mortality risk in humans [12, 13] and longevity in rodents [11].

Exercise, Energy Balance and Longevity in Rodents

Since the fundamental cause of obesity is a chronic imbalance in energy intake and expenditure (positive energy balance), body weight and adiposity can be normalized by either restricting caloric intake, increasing energy expenditure via physical activity, or both to create a negative energy balance. Indeed, exercise can mimic many of the phenotypic changes that occur with CR and oppose many of the hazards attributed to sedentary behavior in rodents and humans. Therefore, it would seem intuitive that exercise-related energy expenditure could be a mimetic for the life-prolonging action of CR. Initial studies on exercise and longevity in rats were conflicting as voluntary running was reported to either shorten, extend or not affect longevity [14]. However, a more recent series of widely cited studies by Holloszy et al. [14–17] in rats discussed below were able to definitively address the exercise and longevity debate. In most of these experiments, Holloszy's group utilized the pathogen-free male Long-Evans rat model that is unique in that it does not increase food intake in response to increased energy expenditure from voluntary wheel running, and hence maintain a lower body weight from exercise. The strength of these studies was their ability to directly compare CR and exercise while controlling for other important confounders such as energy intake by pair-feeding or energy balance by weight matching specific groups over the lifespan.

The main conclusion reached by these studies was that maintaining a lower body weight with exercise (~20–30% reduction) increased mean survival in rats by ~3 months, but unlike CR, failed to extend maximum lifespan [14–17]. However, rats that were calorie restricted (~30%) in order to maintain a similar body weight throughout life as wheel-running rats lived significantly longer on average and had an extension in lifespan. Studies in mice have also found that lifelong voluntary exercise improves mean lifespan but does not extend maximum lifespan [18]. As a caveat, in a study of female Long-Evans rats, which do increase their food intake in response to exercise and maintain a similar body weight as sedentary controls, it was found that exercise improved survival despite no reduction in body weight [19], suggesting that a retardation in growth is not required for exercise to improve survival. A representative depiction of how exercise and CR modulate longevity in rodents is shown in

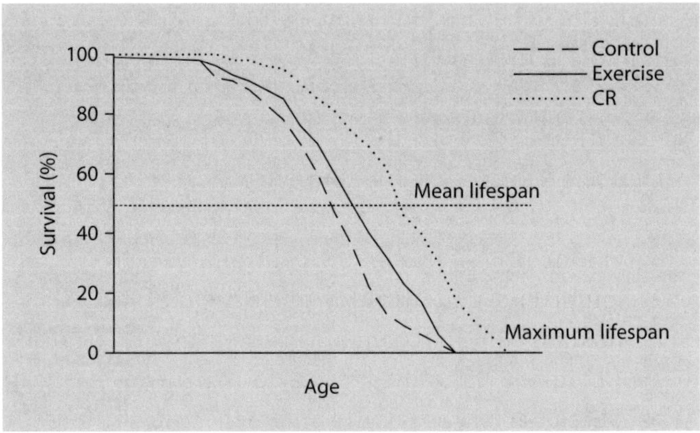

Fig. 1. Representative survival curve of rodents subjected either to sedentary ad libitum-fed conditions, exercise or 30–40% CR. As compared to controls, exercise improves mean lifespan but does not result in lifespan extension. However, CR robustly increases mean lifespan and maximum lifespan in rodents. This suggests that exercise can prevent or delay the development of age-related diseases, but unlike CR, does not slow the fundamental 'rate of aging'.

figure 1. Finally, as will be discussed later, studies in humans [20] have also concluded that exercise confers a survival advantage but does not extend maximum lifespan.

Exercise and Longevity: The Body Fat Paradox

Despite a similar body weight, increasing energy expenditure via exercise in rats fails to mimic the life-prolonging effect of decreased energy intake. However, obesity, and specifically visceral obesity, increases mortality risk in humans [21, 22]. Likewise, as compared to ad libitum-fed controls, surgical removal of visceral fat from male Sprague-Dawley rats resulted in improved mean and maximum lifespan, reduced the hazard rate of death by 51%, and accounted for ~20% of the effect of CR on longevity [11]. Although CR [11] and exercise [23] have been shown to prevent the age-related accrual of visceral fat in rats, several studies have noted that when matched for body weight or body mass index (BMI), exercise results in a leaner phenotype than food restriction. Indeed, the wheel-running rats in the longevity studies by Holloszy and coworkers [14–17, 24] were observed to be leaner than their weight-matched food-restricted sedentary counterparts [25]. This was also observed with lifelong voluntary wheel running in female rats which increased lean mass and attenuated the age-related accrual of body fat at 25 months of age [26]. Similarly, when mice from our study [27], discussed below, were weight matched by exercise or CR for 24 weeks, runners were notably leaner than their sedentary weight-matched counterparts.

Therefore, if reduced adiposity is an important mediator of the beneficial effects of CR, then exercised rodents should be expected to outlive their weight-matched food-restricted controls, but this has proven not to be the case. Interestingly, when matched for BMI, humans engaged in high-volume running are leaner than their calorie-restricted counterparts [28], but whether the same paradox that applies to rodent longevity also applies to humans is unknown.

Disparate Effects of Exercise and Calorie Restriction on Longevity: Potential Mechanisms

In rodents, exercise can promote weight loss and a leaner phenotype than CR, but fails to extend lifespan. Thus, the question remains, why does exercise fail to mimic the life-prolonging action of CR in rodents? One hypothesis is that some harmful byproduct(s) or consequence(s) of exercise at least partially 'interfere' with the beneficial effects of being lean. Certainly, there are many known suspected harmful byproducts of exercise training that could be detrimental for longevity including, increased ROS formation [29], inflammation [30], and heat stress [31]. However, there is currently no good evidence to support an 'interference' effect of exercise. To the contrary, exercise training is known to upregulate antioxidant defenses [32], repair enzymes [33, 34] chaperone proteins [35] and reduce oxidative damage [27, 36]. Furthermore, providing male wheel-running rats with an antioxidant-supplemented diet failed to alter their lifespan [24], suggesting that oxidative damage is not the culprit. In fact, there is now evidence that exercise-induced ROS production is important for cell signaling and the hormetic response to exercise training [37]. For instance, dietary supplementation in humans with antioxidants (vitamins C and E) prevented the insulin-sensitizing effects of exercise training [38].

There is mounting evidence that inflammation is an important modulator of longevity [39] and a single bout of exercise has been shown to cause an acute pro-inflammatory response both systemically and in skeletal muscle [40]. However, exercise training has been reported to have many anti-inflammatory effects [41–43], including reduced inflammation in adipose tissue [44]. On the other hand, acute exercise can transiently raise core body temperature [31] and a reduction in core body temperature is thought to be an important adaptation for CR-mediated longevity [45–47]. However, weight loss in humans by combined CR and exercise was as effective as CR alone at reducing core body temperature [45]. Furthermore, combining exercise with ~30% CR in rats did not abrogate its life-prolonging effect [15], strongly suggesting that exercise does not fail to extend lifespan in mammals due to interference. Interestingly, this may not be universally true among taxa as increased flight activity in drosophila decreases mean and maximum lifespan, which is thought to be due to greater lipid peroxidation levels [48].

A second possibility is that CR extends lifespan by a mechanism(s) independent of leanness per se, and the physiologic response to increased energy expenditure via exercise, fails to completely mimic a reduced flux in nutrients. This has led to several studies probing possibilities beyond changes in body size and adiposity that might distinguish CR from exercise. It is important to keep in mind that identifying the bonafide mechanisms responsible for the disparate effect of CR and exercise on longevity is complicated by the fact that the exact mechanism(s) responsible for the life-prolonging action of CR remain poorly understood. Certainly, numerous hypotheses have been proposed in addition to growth retardation and fat loss [7, 49]. As will be discussed in more detail, some of the biomarkers of aging in which CR and exercise have been directly compared include indicators of endocrine and metabolic health, oxidative stress and inflammatory mediators.

Exercise, Calorie Restriction and Biomarkers of Aging in Humans

Several studies from Fontana and colleagues at Washington University have attempted to characterize the effects of either long-term self-imposed CR or high-volume endurance exercise training on biomarkers of aging and disease risk in humans. In one study [50], thyroid function was assessed and compared in individuals on either self-imposed CR for an average of 6 years (range 3–15 years) versus endurance runners that had engaged in high-volume exercise (~48 miles per week) for an average of 21 years. Exercisers were matched for age, sex, and percentage fat with the CR group. A third group of sedentary individuals consuming a normal western diet were included as controls. The results of this study found that serum triiodothyronine (T_3) was significantly lower in the CR group than the high-volume exercisers or controls. A low serum concentration of T_3 has been observed in rodents [51], humans [45], and monkeys [46] on CR and is suspected to be important for the beneficial effects of restricting food intake. Furthermore, serum levels of the pro-inflammatory cytokine TNF-α were significantly lower in the CR group than in exercisers or controls. Taken together, these results suggest that CR, but not exercise, lowers serum T_3 and TNF-α levels, consistent with changes observed in animals on CR.

In a follow-up investigation [28], individuals on a low-calorie, low-protein diet with high-volume runners matched for BMI and gender and controls consuming a western diet were compared for cancer risk factors. A particular focus was placed on the IGF-1 axis since low IGF-1 levels or action have been associated with less cancer risk [52] and improved longevity [53–55]. They observed that the low-calorie, low protein group had significantly lower serum levels of IGF-1, IGFBP-3, and IGF-1 to IGFBP-3 ratio along with greater sex-hormone-binding globulin than runners or controls [28]. Although exercise resulted in many favorable changes as compared to controls, including lower levels of serum insulin and inflammatory markers, the results of this study suggested that a low-calorie, low-protein diet may confer added protection from cancer risk. In a third study [56], they assessed differences in CVD risk factors among individuals on a self-imposed low-calorie vegan diet, high-volume exercisers matched for BMI, and controls. Both high-volume runners and the low-calorie vegan group had a more favorable CVD risk profile than controls. Interestingly, the low-calorie vegan group had lower systolic and diastolic blood pressure than the exercisers suggesting that the vegan diet provided an additional benefit for cardiovascular health.

A more recent study in this same population reported that CR and exercise were associated with improved insulin sensitivity as compared to sedentary controls [25]. This was accompanied in the CR group by greater plasma adiponectin levels and lower resistin levels than exercisers or controls. However, individuals on CR were found to have impaired glucose tolerance by 90 and 120 min after carbohydrate ingestion and this was accompanied by higher insulin and C-peptide levels. Meanwhile, exercisers were found to have the most favorable response to the glucose load including a better glucose excursion and generally lower C-peptide and insulin response at the later time points (90 and 120 min) [25]. In summary, although several benefits were associated with exercise, these data support that long-term CR is even more effective than exercise for reducing a wide variety of disease risk factors, particularly during the fasting state, but exercise training may confer an advantage for glucose homeostasis in the fed state.

Exercise, Calorie Restriction and Biomarkers of Aging in Rodents

In 1981, Reaven and Reaven [57] performed a study designed to directly compare exercise and CR in rats for endocrine and structural changes to the pancreas with aging. Four-month-old male rats were either assigned to a sedentary group, voluntary exercise group or were calorie-restricted to remain weight-matched with the exercising rats until 12 months of age; 4-month-old rats were also included as young controls [57]. They observed that 12-month-old controls had elevated insulin levels and isolated islets showed significantly reduced glucose-induced insulin release compared with islets from young rats. The 12-month-old exercised or calorie-restricted rats had normal serum insulin levels and did not demonstrate any of the detrimental pathologic changes observed in the pancreas of 12-month-old sedentary controls. Surprisingly, the ability of beta cell islets from the CR group to secrete insulin in vitro in response to increasing glucose concentrations was no better than controls while the authors cautiously concluded that only exercise mildly improved beta cell function [57]. However, a follow-up study from this group found that exercise training or CR were equally effective at preventing hypertriglyceridemia in rats [58]. Meanwhile, exercise training was also found to better protect against the development of adipocyte hypertrophy and insulin resistance than CR in rats [59].

We performed a carefully controlled experiment in mice designed to directly compare CR and exercise on biomarkers relevant to aging [27]. Groups were young male mice (12 wks old) and adult sedentary, caloric-restricted and exercised male mice (36 wks old). Confounders were controlled for by either pair-feeding groups to control for energy intake or weight matching to control for energy balance. Perhaps the most important outcome we observed was that even at the same body weight, mild (~9%) and moderate CR (~18%) were generally more effective at lowering fasting IGF-1 and insulin levels than exercise, while an exercise plus ~9% CR combination reduced levels to a similar extent as weight matched mice on 18% CR [27]. This observation appears to be in line with similar comparisons in humans just described [28] and is consistent with the notion that a reduction in the insulin/IGF-1 axis is an important feature of the life-extending effect of CR.

In rodents, a variety of other physiologic adaptations to CR and exercise with relevance to healthy aging have been studied, with many reporting a similar response to either intervention. For example, both exercise and CR have been found to improve neuronal survival [60], cognitive function [61, 62] and to increase brain-derived neurotrophic factor (BDNF) levels [63], though neither exercise or CR were able to prevent the age-related decline in pituitary-adrenal function in rats [64]. In addition, exercise and CR can increase neurogenesis, but recent studies suggest that the enhancement of neurogenesis may occur by different neurobiological mechanisms [65]. Exercise and CR have been shown to induce mitochondrial biogenesis and reduce oxidative stress in rodents [27, 66, 67]. Likewise, Leeuwenburgh and colleagues have reported that lifelong exercise alone can attenuate markers of oxidative damage in the cerebellum of rats [68]. Furthermore, both 8% lifelong CR and 8% CR plus exercise in rats was able to attenuate age-related changes in the plantaris muscle [69] and the addition of exercise to 8% CR led to a greater reduction in serum CRP [70]. However, exercise appeared to add no further benefit to 8% CR for attenuating catecholamine biosynthetic enzyme levels with age [71] several measures of oxidative stress in liver [72], or in limiting the age-related decline in skeletal muscle autophagy [73].

Unexpectedly, our study in pair-fed and weight-matched mice found that while obesity from a high-fat diet accelerated the accumulation of 8-OHdG levels in liver, only heavy exercise completely attenuated the age-related accrual of liver 8-OHdG [27], a finding that is consistent with reports that exercise improves DNA repair capacity [32]. Therefore, we concluded that the failure of exercise to extend lifespan may be due to an inability to fully mimic the hormonal or metabolic response to CR, including an inability to as effectively lower serum insulin and IGF-1 levels, rather than an oxidative damage-mediated interference. Collectively, data in rodents demonstrate that while both exercise and CR can improve a wide array of parameters indicative of healthy aging and longevity, there are also important differences. The relevance of these disparate effects and overlap to longevity in rodents, and more importantly humans, remains to be elucidated.

Effect of Exercise and Diet-Induced Weight Loss on Disease Risk in Humans

Several studies have evaluated the effect of exercise and diet-induced weight loss on changes in disease risk factors. A 12-month intervention study in overweight men and women utilizing either 20% CR or an exercise program designed to increase energy expenditure by 20%, resulted in similar reductions in body weight and body fat, and similar improvements in CVD risk factors [74]. Likewise, it was found that either 20% CR or exercise were similarly effective at reducing markers of DNA and RNA damage to white blood cells [75]. Unlike long-term CR and exercise, a 1-year intervention of either exercise training or CR, which resulted in a comparable energy deficit and weight loss in overweight men and women, led to similar improvements in insulin sensitivity and glucose tolerance [76] but only CR-induced weight loss resulted in a reduced serum T_3 levels [77]. In the CALERIE study from the Pennington group [45, 78], it was reported that 6 months of 25% CR or CR plus exercise combination (12.5% CR, 12.5% increase in energy expenditure) to induce ~10% weight loss were equally effective at reducing fasting insulin, core body temperature, DNA damage and stimulating mitochondrial biogenesis in skeletal muscle [45]. Thus, these studies suggest that either diet, exercise or a diet plus exercise combination generally produce similar health benefits with comparable amounts of weight loss in overweight individuals.

Other studies from these same groups however have identified parameters where weight loss by exercise or a combination of diet and exercise is even more beneficial than CR alone. For instance, while both CR and CR plus exercise for 12 months resulted in similar losses of body weight, total body fat and visceral fat [79], only CR plus exercise resulted in enhanced insulin sensitivity and a greater reduction in LDL cholesterol and diastolic blood pressure. Furthermore, the CR plus exercise group saw a significant improvement in aerobic capacity while the CR only group saw a slight deterioration in aerobic function [79]. Likewise, the same group reported a larger reduction in 10-year CVD risk with weight loss when exercise was included [80] which taken together, suggests that exercise provides benefits beyond what can be achieved by weight and fat loss alone. Furthermore, unlike CR, the inclusion of exercise during weight loss was better at preserving bone density [81] and improving lean muscle mass, strength and aerobic capacity [82], whereas these parameters were reduced in the CR only group. Thus, not all aspects of a CR regimen to achieve weight loss are beneficial in humans, and incorporating exercise with CR may present a more viable strategy for improving health and function in both the general populations as well as specific subgroups such as frail and elderly individuals.

Exercise and Longevity in Humans

Exercise and Mortality Risk in Humans

It has long been suspected that engaging in regular physical activity is beneficial for human longevity. Indeed, exercise has been shown to reduce the risk for CVD [83], stroke [84], T2DM [85], osteoporosis [86], and sarcopenia in humans [87]. In addition, exercise is associated with reduced risk of developing or dying from certain cancers including, breast [88], colon [89], and prostate [90–93], helps to preserve cognitive function [94], and minimizes the risk of developing Alzheimer's disease [95]. However, studies attempting to demonstrate a relationship between exercise and mortality risk in humans are complicated by several nuances such as the importance of exercise intensity, energy expenditure, cumulative physical activity and the reliability of studies using self-report to classify activity levels, rather than a more quantitative measure.

The earliest studies interested in the relationship between physical activity and morbidity assessed the rate of CVD in London workers of various occupations [96] and determined that increased activity levels (work, leisure) were associated with lower rates of CVD. Likewise, the widely cited Harvard Alumni Study from Lee et al. [97] confirmed that expending 2,000 kcal or more per week in daily physical activity added ~2 years of life in men but with little added benefit beyond >3,500 kcal expended per week. Similarly, the Framingham Heart Study found that increased physical activity in women reduces all-cause mortality risk in a dose-response manner [98]. More recently, the Jerusalem Longitudinal Cohort Study reported a significant survival benefit when physical activity was initiated between the ages of 70–78 or 78–85 years and was associated with a higher likelihood of remaining independent while performing daily activities at age 85 years [99]. In a prospective study of 2,357 healthy men, regular exercise reduced the mortality risk by nearly 30%, and in the absence of several risk factors including obesity and a sedentary lifestyle, resulted in a 54% probability of a 90-year lifespan at age 70 years [100].

Role of Exercise Intensity versus Quantity

Though the aforementioned studies clearly demonstrate that exercise is effective at reducing mortality risk, they do not necessary address the issue of whether vigorous or nonvigorous activity is most beneficial. Two follow-up reports from the Harvard Alumni Study [97, 101] found that either moderate or vigorous activities, such as climbing stairs, but not light activities were associated with improved survival. Similarly, the US Railroad Study found that improved fitness was associated with lower risk of death, and remarkably found that the benefit of 100 kcal/week expended in vigorous activities offered equivalent protection from CVD risk as 1,000 kcal/week by light-to-moderate activity [102]. There is evidence that elite endurance athletes and mixed sports athletes have increased survival rates as compared to the general population [103] and men that were regular joggers had significantly lower mortality rates than nonjoggers [104]. Furthermore, performance on a 400-meter walking test in Italian men and women ≥65 years was inversely related to mortality risk [105]. Interestingly, data from the Jerusalem Longitudinal Cohort Study indicated that the beneficial effects of exercise were sufficiently achieved with a moderate level of activity as no added benefit was observed with higher levels of activity [99]. Thus, while the importance of participating in vigorous exercise to lower mortality risk is not entirely clear, most studies seem to favor moderate-to-vigorous activity over leisure activities.

An important limitation of the aforementioned studies on exercise intensity is their reliance on self report of activity levels and habits and some have suggested that this may have resulted in underestimating the importance of cumulative physical activity. In an effort to objectively measure activity level, Manini et al. [106] utilized the doubly labeled water technique as a quantitative measure of activity-related energy expenditure. Remarkably, they observed that a difference of ~287 kcal/day in activity-related energy expenditure (exercise and nonexercise related) was associated with a 32% lower risk of mortality after adjusting for several variables. Furthermore, a dose response was observed between activity-related energy expenditure and mortality such that an absolute risk of death was 12.1% in the highest tertile 17.6% in the middle, and 24.7% in the lowest tertile. The significance of this study was that it challenges the conclusions of prior studies and supports the hypothesis that cumulative energy expenditure that includes leisure activities, irrespective of exercise intensity, is associated with improved survival in older adults.

Exercise and Maximum Lifespan

Despite numerous studies demonstrating that both accumulated and moderate to vigorous physical activity prevents premature mortality [106–108], the author is only aware of a single study examining the impact of exercise on maximum lifespan in humans. In this study, men from Finland were prospectively followed over 20 years [20]. They observed that over the first two-thirds of follow-up, men engaged in high amounts of physical activity had improved survival as compared to men classified to a low-exercise group [20]. However, during the last third, the survival curves converged between these two groups, demonstrating no difference in maximum lifespan. Thus, this finding is in agreement with the conclusions drawn from rodents that exercise improves mean lifespan by preventing premature death, but does not fundamentally slow the rate of aging.

Fitness versus Fatness

Though it is clear that increased physical activity can reduce the risk of death in humans, most of these studies just described did not account or adjust for important confounders including differences in dietary intake, body weight or body composition. Indeed, there has been great interest over the years in whether fitness and obesity independently contribute to cardiometabolic risk and mortality or if the effects of exercise require significant changes in body composition. More importantly, as the risks posed by obesity to disease and mortality are well documented, can high cardiorespiratory fitness abrogate the risk posed by overweight and obesity? Overall, the data have been equivocal. Numerous studies from the Cooper Clinic and others [109, 110] have suggested that a high level of fitness can completely attenuate the risk posed by obesity, as classified by BMI, for CVD [111], hypertension [112], metabolic syndrome [108] and mortality [113]. Others have concluded that both fitness and obesity are independent contributors to morbidity and mortality from CVD and all causes such that fitness should not be expected to completely protect from the hazards posed by obesity [114, 115].

Meanwhile, most but not all the weight loss studies described have reported greater benefit for disease risk factors when similar weight loss is achieved by exercise or combining CR and exercise, rather than CR alone. Thus, there is good evidence in humans that exercise-mediated weight loss confers protection from disease risk beyond leanness achieved by diet [79]. Furthermore, unlike CR, incorporating exercise with diet-induced weight loss preserves bone density [81] and improves lean muscle mass, strength and aerobic capacity [82]. Thus, the available evidence seems to favor that an exercise component be included in a weight loss intervention, in order to maximize the health benefits of weight loss to promote healthy aging.

Theoretical Considerations

Despite the fact that rodent studies convincingly show greater benefit from CR than exercise [15], and that exercise fails to extend lifespan in humans [20], there is reason to be more optimistic about the benefits of exercise, rather than CR for promoting healthy aging in humans. First of all, several of the intervention studies discussed have demonstrated similar benefits from CR and exercise, with exercise demonstrating a better propensity than CR to improve bone health [81] and physical function [82]. Additionally, as pointed out previously [116], an important distinction to be made between rodents and humans is how they die. A majority of humans die of cardiovascular complications including CVD and stroke, but this is a very minor cause of death in mice and rats, whose deaths are most often attributed to cancers and renal disease. Since exercise is a well-known strategy for preventing cardiometabolic diseases [84, 117], it is plausible that the benefits of exercise in humans are more impressive than in rodents and more closely resemble those observed with CR. Thus, while there is no evidence to support the notion that exercise can extend human lifespan, there are plenty of data to support that exercise combined with maintaining a healthy weight can substantially improve health aging and prevent or delay the onset of the most common causes of death in humans, such as CVD, T2DM and stroke.

Second, while longevity studies in rodents favor CR over exercise for improving mean and maximum lifespan [15, 16], it is unknown if this relationship holds in humans since performing well-controlled longevity studies are not feasible. For this reason, some hope to extrapolate from nonhuman primate studies whether CR may extend lifespan in humans [118]. While these studies seem to show a survival advantage with CR [119], they are not yet far enough along to evaluate maximum lifespan. There is evidence described from the studies by Fontana et al. [28, 50, 120] that long-term CR, but not exercise in humans, leads to many of the same changes in the hormonal milieu observed in rodents and monkeys on CR that are thought to be important for slowing the rate of aging.

Perhaps the best glimpse of how CR may affect human longevity currently comes from observation studies on the natives of Okinawa, Japan who have the highest frequency of centenarians in the world. Okinawans have an average calorie intake of ~1,785 kcal/day over their lifespan which is considerably less than the rest of the developed world, and nearly 17% less than the rest of Japan [121]. As this chapter has given particular focus to the role of body composition, it is important to point out that Okinawans are also very lean with a BMI of ~21, and do not demonstrate any substantial weight gain after peak body weight is reached in early adulthood. However, there are certainly other possible contributors to their exceptional longevity that must be considered

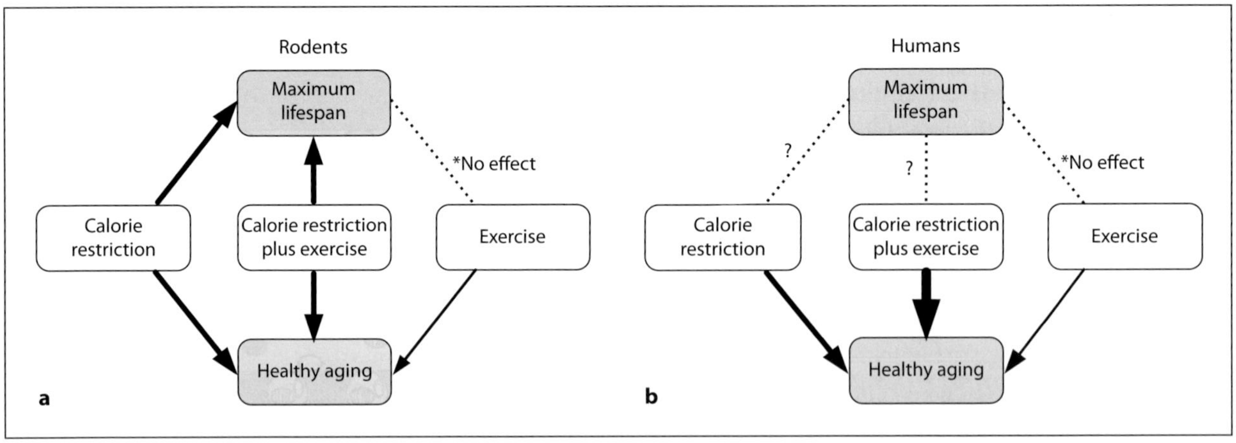

Fig. 2. Schematic diagram representing the impact of CR, exercise and CR plus exercise on healthy aging and longevity in rodents and humans. **a** In rodents, CR can robustly improve survival, healthy aging and result in lifespan extension. In contrast, exercise modestly improves survival in rodents but does not extend lifespan. The combination of exercise plus CR results in a similar improvement in mean and maximum lifespan as CR alone. There is some evidence that further benefit can be achieved by addition of exercise to mild CR (8–18%), but it is unclear if this occurs under more robust CR conditions (30–40%). **b** In humans, comparing CR and exercise is more complex. Studies in individuals engaged in long-term CR or exercise reveal that while both lifestyles are beneficial, further benefits associated with longevity in rodents are observed with CR or its variations (vegan diet, low-protein diet). However, short-term interventions seem to suggest that exercise or CR resulting in comparable weight loss, generally lead to similar improvements in biomarkers of disease risk and aging. Interestingly, similar studies have revealed that CR and exercise in combination is associated with greater benefit for several indicators of health and function than CR alone. Furthermore, limited evidence in humans suggests that exercise does not extend human lifespan, while the effect of CR or CR plus exercise is unknown. Solid arrows indicate a direct effect of an intervention and the line/arrow size is indicative of the relationship strength. Dotted lines indicate either no effect or unknown effect.

including, underlying genetic factors and their interaction with the environment, a diet high in fiber, antioxidants, and low in refined sugars, and a physically active lifestyle. Furthermore, while they have been reported to have the greatest mean life expectancy, there is no evidence of maximum lifespan extension in this population.

Conclusions

It was initially postulated that maintaining a lower body weight with exercise might mimic the life-prolonging action of CR. However, longevity studies in rodents have revealed that exercise only modestly improves survival while failing to extend maximum lifespan. This is in contrast to the robust increase in mean and maximum lifespan observed with CR. As exercise results in a leaner phenotype than CR, several possibilities have been examined beyond changes in body size and adiposity that might distinguish CR from exercise. Evidence suggests that the failure of exercise to extend lifespan in rodents is due to an inability to completely mimic the physiologic response to a reduction in energy intake rather than interference from potentially harmful byproduct(s) of exercise. It is worth speculating that the lack of interference from exercise may be isolated to its interaction with CR as surgical removal of visceral fat under ad libitum-fed conditions, extends lifespan, yet a reduction in adiposity, and presumably visceral fat by exercise does not extend lifespan.

In humans, there is somewhat conflicting evidence as to whether CR or exercise is better for healthy aging and longevity. When matched for BMI, the metabolic profile from long-term CR seems to be slightly more favorable than that from high-volume exercise (48 miles per week), though it is unknown if this amount of exercise was optimal or excessive. In contrast, short-term weight loss by exercise and CR produce similar benefits and, in some instances, favor exercise over CR. As exercise is associated with certain benefits important for health and function not observed with CR, a strategy combining CR and exercise may be optimal for extending health span in humans. The relationship of these strategies with healthy aging and longevity in rodents and humans is depicted in figure 2. Whether either strategy, alone or in combination, is capable of achieving true lifespan extension in humans remains to be resolved.

Acknowledgements

This work was supported by the University of Alabama at Birmingham (UAB) Center for Aging Pilot/Feasibility Grant Program, the UAB Clinical Nutrition Research Unit (P30 DK56336), and an NCI Sponsored Training Grant (CA 47888). D.M.H. is currently supported by a postdoctoral T32 Training Grant at the Albert Einstein College of Medicine (T32 AG23475) and a postdoctoral Fellowship Award from the Prevent Cancer Foundation.

References

1 Weindruch R, Walford R: The Retardation of Aging and Disease by Dietary Restriction. New York, Thomas, 1988.
2 Tannenbaum A: The dependence of the genesis of induced skin tumors on the caloric intake during different stages of carcinogenesis. Cancer Res 1944; 4:673–679.
3 Kealy RD, Lawler DF, Ballam JM, Mantz SL, Biery DN, Greeley EH, Lust G, Segre M, Smith GK, Stowe HD: Effects of diet restriction on life span and age-related changes in dogs. J Am Vet Med Assoc 2002; 220:1315–1320.
4 Roth GS, Ingram DK, Lane MA: Caloric restriction in primates and relevance to humans. Ann NY Acad Sci 2001;928:305–315.
5 Pinney DO, Stephens DF, Pope LS: Lifetime effects of winter supplemental feed level and age at first parturition on range beef cows. J Anim Sci 1972;34: 1067–1074.
6 Pugh TD, Klopp RG, Weindruch R: Controlling caloric consumption: protocols for rodents and rhesus monkeys. Neurobiol Aging 1999;20:157–165.
7 Masoro EJ: Caloric restriction-induced life extension of rats and mice: a critique of proposed mechanisms. Biochim Biophys Acta 2009;1790:1040–1048.
8 Barzilai N, Gupta G: Revisiting the role of fat mass in the life extension induced by caloric restriction. J Gerontol A Biol Sci Med Sci 1999;54:B89–B96; discussion B97–B98.

9 Michell AR: Longevity of British breeds of dog and its relationships with sex, size, cardiovascular variables and disease. Vet Rec 1999;145:625–629.
10 Bartke A: Insulin and aging. Cell Cycle 2008;7:3338–3343.
11 Muzumdar R, Allison DB, Huffman DM, Ma X, Atzmon G, Einstein FH, Fishman S, Poduval AD, McVei T, Keith SW, Barzilai N: Visceral adipose tissue modulates mammalian longevity. Aging Cell 2008;7:438–440.
12 Walker SP, Rimm EB, Ascherio A, Kawachi I, Stampfer MJ, Willett WC: Body size and fat distribution as predictors of stroke among US men. Am J Epidemiol 1996;144:1143–1150.
13 Pischon T, et al: General and abdominal adiposity and risk of death in Europe. N Engl J Med 2008;359: 2105–2120.
14 Holloszy JO, Smith EK: Effects of exercise on longevity of rats. Fed Proc 1987;46:1850–1853.
15 Holloszy JO: Mortality rate and longevity of food-restricted exercising male rats: a reevaluation. J Appl Physiol 1997;82:399–403.
16 Holloszy JO, Smith EK, Vining M, Adams S: Effect of voluntary exercise on longevity of rats. J Appl Physiol 1985;59:826–831.
17 Holloszy JO, Schechtman KB: Interaction between exercise and food restriction: effects on longevity of male rats. J Appl Physiol 1991;70:1529–1535.
18 Samorajski T, Delaney C, Durham L, Ordy JM, Johnson JA, Dunlap WP: Effect of exercise on longevity, body weight, locomotor performance, and passive-avoidance memory of C57BL/6J mice. Neurobiol Aging 1985;6:17–24.
19 Holloszy JO: Exercise increases average longevity of female rats despite increased food intake and no growth retardation. J Gerontol 1993;48:B97–B100.
20 Pekkanen J, Marti B, Nissinen A, Tuomilehto J, Punsar S, Karvonen MJ: Reduction of premature mortality by high physical activity: a 20-year follow-up of middle-aged Finnish men. Lancet 1987;i:1473–1477.
21 Postorino M, Marino C, Tripepi G, Zoccali C: Abdominal obesity and all-cause and cardiovascular mortality in end-stage renal disease. J Am Coll Cardiol 2009;53:1265–1272.
22 Price GM, Uauy R, Breeze E, Bulpitt CJ, Fletcher AE: Weight, shape, and mortality risk in older persons: elevated waist-hip ratio, not high body mass index, is associated with a greater risk of death. Am J Clin Nutr 2006;84:449–460.
23 Gollisch KS, Brandauer J, Jessen N, Toyoda T, Nayer A, Hirshman MF, Goodyear LJ: Effects of exercise training on subcutaneous and visceral adipose tissue in normal- and high-fat diet-fed rats. Am J Physiol Endocrinol Metab 2009;297:E495–E504.
24 Holloszy JO: Longevity of exercising male rats: effect of an antioxidant supplemented diet. Mech Ageing Dev 1998;100:211–219.
25 Fontana L, Klein S, Holloszy JO: Effects of long-term calorie restriction and endurance exercise on glucose tolerance, insulin action, and adipokine production. Age (Dordr) 2010;32:97–108.
26 Slentz CA, Holloszy JO: Body composition of physically inactive and active 25-month-old female rats. Mech Age Dev 1993;69:161–166.
27 Huffman DM, Moellering DR, Grizzle WE, Stockard CR, Johnson MS, Nagy TR: Effect of exercise and calorie restriction on biomarkers of aging in mice. Am J Physiol Regul Integr Comp Physiol 2008;294: R1618–R1627.
28 Fontana L, Klein S, Holloszy JO: Long-term low-protein, low-calorie diet and endurance exercise modulate metabolic factors associated with cancer risk. Am J Clin Nutr 2006;84:1456–1462.
29 Ji LL: Antioxidants and oxidative stress in exercise. Proc Soc Exp Biol Med 1999;222:283–292.
30 Chung HY, Kim HJ, Kim JW, Yu BP: The inflammation hypothesis of aging: molecular modulation by calorie restriction. Ann NY Acad Sci 2001;928:327–335.
31 Jay O, Reardon FD, Webb P, Ducharme MB, Ramsay T, Nettlefold L, Kenny GP: Estimating changes in mean body temperature for humans during exercise using core and skin temperatures is inaccurate even with a correction factor. J Appl Physiol 2007;103: 443–451.
32 Goto S, Naito H, Kaneko T, Chung HY, Radak Z: Hormetic effects of regular exercise in aging: correlation with oxidative stress. Appl Physiol Nutr Metab 2007;32:948–953.
33 Ji LL, Gomez-Cabrera MC, Vina J: Exercise and hormesis: activation of cellular antioxidant signaling pathway. Ann NY Acad Sci 2006;1067:425–435.
34 Ji LL: Exercise-induced modulation of antioxidant defense. Ann NY Acad Sci 2002;959:82–92.
35 Morton JP, MacLaren DP, Cable NT, Bongers T, Griffiths RD, Campbell IT, Evans L, Kayani A, McArdle A, Drust B: Time course and differential responses of the major heat shock protein families in human skeletal muscle following acute nondamaging treadmill exercise. J Appl Physiol 2006;101: 176–182.
36 Radak Z, Chung HY, Goto S: Exercise and hormesis: oxidative stress-related adaptation for successful aging. Biogerontology 2005;6:71–75.
37 Radak Z, Chung HY, Koltai E, Taylor AW, Goto S: Exercise, oxidative stress and hormesis. Ageing Res Rev 2008;7:34–42.

38 Ristow M, Zarse K, Oberbach A, Kloting N, Birringer M, Kiehntopf M, Stumvoll M, Kahn CR, Bluher M: Antioxidants prevent health-promoting effects of physical exercise in humans. Proc Natl Acad Sci USA 2009;106:8665–8670.
39 Chung HY, Cesari M, Anton S, Marzetti E, Giovannini S, Seo AY, Carter C, Yu BP, Leeuwenburgh C: Molecular inflammation: underpinnings of aging and age-related diseases. Ageing Res Rev 2009;8:18–30.
40 Pedersen BK, Ostrowski K, Rohde T, Bruunsgaard H: The cytokine response to strenuous exercise. Can J Physiol Pharmacol 1998;76:505–511.
41 Nicklas BJ, Hsu FC, Brinkley TJ, Church T, Goodpaster BH, Kritchevsky SB, Pahor M: Exercise training and plasma C-reactive protein and interleukin-6 in elderly people. J Am Geriatr Soc 2008;56:2045–2052.
42 Silverman NE, Nicklas BJ, Ryan AS: Addition of aerobic exercise to a weight loss program increases BMD, with an associated reduction in inflammation in overweight postmenopausal women. Calcif Tissue Int 2009;84:257–265.
43 Petersen AM, Pedersen BK: The role of IL-6 in mediating the anti-inflammatory effects of exercise. J Physiol Pharmacol 2006;57(suppl 10):43–51.
44 Vieira VJ, Valentine RJ, Wilund KR, Antao N, Baynard T, Woods JA: Effects of exercise and low-fat diet on adipose tissue inflammation and metabolic complications in obese mice. Am J Physiol Endocrinol Metab 2009;296:E1164–E1171.
45 Heilbronn LK, de Jonge L, Frisard MI, DeLany JP, Larson-Meyer DE, Rood J, Nguyen T, Martin CK, Volaufova J, Most MM, Greenway FL, Smith SR, Deutsch WA, Williamson DA, Ravussin E: Effect of 6-month calorie restriction on biomarkers of longevity, metabolic adaptation, and oxidative stress in overweight individuals: a randomized controlled trial. JAMA 2006;295:1539–1548.
46 Mattison JA, Lane MA, Roth GS, Ingram DK: Calorie restriction in rhesus monkeys. Exp Gerontol 2003;38:35–46.
47 Lane MA, Baer DJ, Rumpler WV, Weindruch R, Ingram DK, Tilmont EM, Cutler RG, Roth GS: Calorie restriction lowers body temperature in rhesus monkeys, consistent with a postulated anti-aging mechanism in rodents. Proc Natl Acad Sci USA 1996;93:4159–4164.
48 Magwere T, Pamplona R, Miwa S, Martinez-Diaz P, Portero-Otin M, Brand MD, Partridge L: Flight activity, mortality rates, and lipoxidative damage in Drosophila. J Gerontol A Biol Sci Med Sci 2006;61: 136–145.
49 Barja G: Rate of generation of oxidative stress-related damage and animal longevity. Free Radic Biol Med 2002;33:1167–1172.
50 Fontana L, Klein S, Holloszy JO, Premachandra BN: Effect of long-term calorie restriction with adequate protein and micronutrients on thyroid hormones. J Clin Endocrinol Metab 2006;91:3232–3235.
51 Rikke BA, Yerg JE 3rd, Battaglia ME, Nagy TR, Allison DB, Johnson TE: Strain variation in the response of body temperature to dietary restriction. Mech Ageing Dev 2003;124:663–678.
52 Kaaks R, Toniolo P, Akhmedkhanov A, Lukanova A, Biessy C, Dechaud H, Rinaldi S, Zeleniuch-Jacquotte A, Shore RE, Riboli E: Serum C-peptide, insulin-like growth factor (IGF)-I, IGF-binding proteins, and colorectal cancer risk in women. J Natl Cancer Inst 2000;92:1592–1600.
53 Yuan R, Tsaih SW, Petkova SB, Marin de Evsikova C, Xing S, Marion MA, Bogue MA, Mills KD, Peters LL, Bult CJ, Rosen CJ, Sundberg JP, Harrison DE, Churchill GA, Paigen B: Aging in inbred strains of mice: study design and interim report on median lifespans and circulating IGF1 levels. Aging Cell 2009;8:277–287.
54 Kappeler L, De Magalhaes Filho C, Dupont J, Leneuve P, Cervera P, Perin L, Loudes C, Blaise A, Klein R, Epelbaum J, Le Bouc Y, Holzenberger M: Brain IGF-1 receptors control mammalian growth and lifespan through a neuroendocrine mechanism. PLoS Biol 2008;6:e254.
55 Suh Y, Atzmon G, Cho MO, Hwang D, Liu B, Leahy DJ, Barzilai N, Cohen P: Functionally significant insulin-like growth factor I receptor mutations in centenarians. Proc Natl Acad Sci USA 2008;105: 3438–3442.
56 Fontana L, Meyer TE, Klein S, Holloszy JO: Long-term low-calorie low-protein vegan diet and endurance exercise are associated with low cardiometabolic risk. Rejuvenation Res 2007;10:225–234.
57 Reaven EP, Reaven GM: Structure and function changes in the endocrine pancreas of aging rats with reference to the modulating effects of exercise and caloric restriction. J Clin Invest 1981;68:75–84.
58 Reaven GM, Reaven EP: Prevention of age-related hypertriglyceridemia by caloric restriction and exercise training in the rat. Metabolism 1981;30:982–986.
59 Craig BW, Garthwaite SM, Holloszy JO: Adipocyte insulin resistance: effects of aging, obesity, exercise, and food restriction. J Appl Physiol 1987;62:95–100.
60 Mattson MP: Neuroprotective signaling and the aging brain: take away my food and let me run. Brain Res 2000;886:47–53.
61 Halagappa VK, Guo Z, Pearson M, Matsuoka Y, Cutler RG, Laferla FM, Mattson MP: Intermittent fasting and caloric restriction ameliorate age-related behavioral deficits in the triple-transgenic mouse model of Alzheimer's disease. Neurobiol Dis 2007; 26:212–220.

62 Stranahan AM, Lee K, Becker KG, Zhang Y, Maudsley S, Martin B, Cutler RG, Mattson MP: Hippocampal gene expression patterns underlying the enhancement of memory by running in aged mice. Neurobiol Aging 2008 [Epub ahead of print].
63 Stranahan AM, Lee K, Martin B, Maudsley S, Golden E, Cutler RG, Mattson MP: Voluntary exercise and caloric restriction enhance hippocampal dendritic spine density and BDNF levels in diabetic mice. Hippocampus 2009;19:951–961.
64 Dellwo M, Beauchene RE: The effect of exercise, diet restriction, and aging on the pituitary–adrenal axis in the rat. Exp Gerontol 1990;25:553–562.
65 Levenson CW, Rich NJ: Eat less, live longer? New insights into the role of caloric restriction in the brain. Nutr Rev 2007;65:412–415.
66 Nisoli E, Tonello C, Cardile A, Cozzi V, Bracale R, Tedesco L, Falcone S, Valerio A, Cantoni O, Clementi E, Moncada S, Carruba MO: Calorie restriction promotes mitochondrial biogenesis by inducing the expression of eNOS. Science 2005;310:314–317.
67 Sohal RS, Agarwal S, Candas M, Forster MJ, Lal H: Effect of age and caloric restriction on DNA oxidative damage in different tissues of C57BL/6 mice. Mech Ageing Dev 1994;76:215–224.
68 Cui L, Hofer T, Rani A, Leeuwenburgh C, Foster TC: Comparison of lifelong and late life exercise on oxidative stress in the cerebellum. Neurobiol Aging 2009;30:903–909.
69 Kim JH, Kwak HB, Leeuwenburgh C, Lawler JM: Lifelong exercise and mild (8%) caloric restriction attenuate age-induced alterations in plantaris muscle morphology, oxidative stress and IGF-1 in the Fischer-344 rat. Exp Gerontol 2008;43:317–329.
70 Kalani R, Judge S, Carter C, Pahor M, Leeuwenburgh C: Effects of caloric restriction and exercise on age-related, chronic inflammation assessed by C-reactive protein and interleukin-6. J Gerontol A Biol Sci Med Sci 2006;61:211–217.
71 Erdos B, Broxson CS, Landa T, Scarpace PJ, Leeuwenburgh C, Zhang Y, Tumer N: Effects of life-long caloric restriction and voluntary exercise on age-related changes in levels of catecholamine biosynthetic enzymes and angiotensin II receptors in the rat adrenal medulla and hypothalamus. Exp Gerontol 2007;42:745–752.
72 Seo AY, Hofer T, Sung B, Judge S, Chung HY, Leeuwenburgh C: Hepatic oxidative stress during aging: effects of 8% long-term calorie restriction and lifelong exercise. Antioxid Redox Signal 2006; 8:529–538.
73 Wohlgemuth SE, Seo AY, Marzetti E, Lees HA, Leeuwenburgh C: Skeletal muscle autophagy and apoptosis during aging: effects of calorie restriction and life-long exercise. Exp Gerontol 2010;45:138–148.
74 Fontana L, Villareal DT, Weiss EP, Racette SB, Steger-May K, Klein S, Holloszy JO: Calorie restriction or exercise: effects on coronary heart disease risk factors. A randomized controlled trial. Am J Physiol Endocrinol Metab 2007;293:E197–E202.
75 Hofer T, Fontana L, Anton SD, Weiss EP, Villareal D, Malayappan B, Leeuwenburgh C: Long-term effects of caloric restriction or exercise on DNA and RNA oxidation levels in white blood cells and urine in humans. Rejuvenation Res 2008;11:793–799.
76 Weiss EP, Racette SB, Villareal DT, Fontana L, Steger-May K, Schechtman KB, Klein S, Holloszy JO: Improvements in glucose tolerance and insulin action induced by increasing energy expenditure or decreasing energy intake: a randomized controlled trial. Am J Clin Nutr 2006;84:1033–1042.
77 Weiss EP, Villareal DT, Racette SB, Steger-May K, Premachandra BN, Klein S, Fontana L: Caloric restriction but not exercise-induced reductions in fat mass decrease plasma triiodothyronine concentrations: a randomized controlled trial. Rejuvenation Res 2008;11:605–609.
78 Civitarese AE, Carling S, Heilbronn LK, Hulver MH, Ukropcova B, Deutsch WA, Smith SR, Ravussin E: Calorie restriction increases muscle mitochondrial biogenesis in healthy humans. PLoS Med 2007; 4:e76.
79 Larson-Meyer DE, Redman L, Heilbronn LK, Martin CK, Ravussin E: Caloric restriction with or without exercise: the fitness versus fatness debate. Med Sci Sports Exerc 2010;42:152–159.
80 Lefevre M, Redman LM, Heilbronn LK, Smith JV, Martin CK, Rood JC, Greenway FL, Williamson DA, Smith SR, Ravussin E: Caloric restriction alone and with exercise improves CVD risk in healthy non-obese individuals. Atherosclerosis 2009;203: 206–213.
81 Villareal DT, Fontana L, Weiss EP, Racette SB, Steger-May K, Schechtman KB, Klein S, Holloszy JO: Bone mineral density response to caloric restriction-induced weight loss or exercise-induced weight loss: a randomized controlled trial. Arch Intern Med 2006;166:2502–2510.
82 Weiss EP, Racette SB, Villareal DT, Fontana L, Steger-May K, Schechtman KB, Klein S, Ehsani AA, Holloszy JO: Lower extremity muscle size and strength and aerobic capacity decrease with caloric restriction but not with exercise-induced weight loss. J Appl Physiol 2007;102:634–640.
83 Laughlin MH: Cardiovascular response to exercise. Am J Physiol 1999;277:S244–S259.
84 Wendel-Vos GC, Schuit AJ, Feskens EJ, Boshuizen HC, Verschuren WM, Saris WH, Kromhout D: Physical activity and stroke. A meta-analysis of observational data. Int J Epidemiol 2004;33:787–798.

85 Laaksonen DE, et al: Physical activity in the prevention of type 2 diabetes: the Finnish diabetes prevention study. Diabetes 2005;54:158–165.
86 Vainionpaa A, Korpelainen R, Leppaluoto J, Jamsa T: Effects of high-impact exercise on bone mineral density: a randomized controlled trial in premenopausal women. Osteoporos Int 2005;16:191–197.
87 Roth SM, Ferrell RF, Hurley BF: Strength training for the prevention and treatment of sarcopenia. J Nutr Health Aging 2000;4:143–155.
88 Friedenreich CM: Physical activity and breast cancer risk: the effect of menopausal status. Exerc Sport Sci Rev 2004;32:180–184.
89 Slattery ML, Edwards S, Curtin K, Ma K, Edwards R, Holubkov R, Schaffer D: Physical activity and colorectal cancer. Am J Epidemiol 2003;158:214–224.
90 Hu FB, Willett WC, Li T, Stampfer MJ, Colditz GA, Manson JE: Adiposity as compared with physical activity in predicting mortality among women. N Engl J Med 2004;351:2694–2703.
91 Wright CE, Zborowski JV, Talbott EO, McHugh-Pemu K, Youk A: Dietary intake, physical activity, and obesity in women with polycystic ovary syndrome. Int J Obes Relat Metab Disord 2004;28:1026–1032.
92 Hannan LM, Leitzmann MF, Lacey JV Jr, Colbert LH, Albanes D, Schatzkin A, Schairer C: Physical activity and risk of ovarian cancer: a prospective cohort study in the United States. Cancer Epidemiol Biomarkers Prev 2004;13:765–770.
93 Alfano CM, Klesges RC, Murray DM, Bowen DJ, McTiernan A, Vander Weg MW, Robinson LA, Cartmel B, Thornquist MD, Barnett M, Goodman GE, Omenn GS: Physical activity in relation to all-site and lung cancer incidence and mortality in current and former smokers. Cancer Epidemiol Biomarkers Prev 2004;13:2233–2241.
94 Lautenschlager NT, Cox KL, Flicker L, Foster JK, van Bockxmeer FM, Xiao J, Greenop KR, Almeida OP: Effect of physical activity on cognitive function in older adults at risk for Alzheimer disease: a randomized trial. JAMA 2008;300:1027–1037.
95 Rolland Y, Pillard F, Klapouszczak A, Reynish E, Thomas D, Andrieu S, Riviere D, Vellas B: Exercise program for nursing home residents with Alzheimer's disease: a 1-year randomized, controlled trial. J Am Geriatr Soc 2007;55:158–165.
96 Morris JN, Heady JA, Raffle PA, Roberts CG, Parks JW: Coronary heart-disease and physical activity of work. Lancet 1953;265:1111–1120.
97 Lee IM, Hsieh CC, Paffenbarger RS Jr: Exercise intensity and longevity in men: The Harvard Alumni Health Study. JAMA 1995;273:1179–1184.
98 Sherman SE, D'Agostino RB, Cobb JL, Kannel WB: Physical activity and mortality in women in the Framingham Heart Study. Am Heart J 1994;128:879–884.
99 Stessman J, Hammerman-Rozenberg R, Cohen A, Ein-Mor E, Jacobs JM: Physical activity, function, and longevity among the very old. Arch Intern Med 2009;169:1476–1483.
100 Yates LB, Djousse L, Kurth T, Buring JE, Gaziano JM: Exceptional longevity in men: modifiable factors associated with survival and function to age 90 years. Arch Intern Med 2008;168:284–290.
101 Lee IM, Paffenbarger RS Jr: Associations of light, moderate, and vigorous intensity physical activity with longevity: The Harvard Alumni Health Study. Am J Epidemiol 2000;151:293–299.
102 Slattery ML, Jacobs DR Jr, Nichaman MZ: Leisure time physical activity and coronary heart disease death: the US Railroad Study. Circulation 1989;79: 304–311.
103 Teramoto M, Bungum TJ: Mortality and longevity of elite athletes. J Sci Med Sport 2009.
104 Schnohr P, Parner J, Lange P: Joggers live longer: The Osterbro study. Ugeskr Laeger 2001;163:2633–2635.
105 Vestergaard S, Patel KV, Bandinelli S, Ferrucci L, Guralnik JM: Characteristics of 400-meter walk test performance and subsequent mortality in older adults. Rejuvenation Res 2009;12:177–184.
106 Manini TM, Everhart JE, Patel KV, Schoeller DA, Colbert LH, Visser M, Tylavsky F, Bauer DC, Goodpaster BH, Harris TB: Daily activity energy expenditure and mortality among older adults. JAMA 2006;296:171–179.
107 Paffenbarger RS Jr, Lee IM: Intensity of physical activity related to incidence of hypertension and all-cause mortality: an epidemiological view. Blood Press Monit 1997;2:115–123.
108 Finley CE, LaMonte MJ, Waslien CI, Barlow CE, Blair SN, Nichaman MZ: Cardiorespiratory fitness, macronutrient intake, and the metabolic syndrome: the Aerobics Center Longitudinal Study. J Am Diet Assoc 2006;106:673–679.
109 Lee CD, Blair SN, Jackson AS: Cardiorespiratory fitness, body composition, and all-cause and cardiovascular disease mortality in men. Am J Clin Nutr 1999;69:373–380.
110 Lee S, Kuk JL, Katzmarzyk PT, Blair SN, Church TS, Ross R: Cardiorespiratory fitness attenuates metabolic risk independent of abdominal subcutaneous and visceral fat in men. Diabetes Care 2005;28:895–901.
111 Blair SN, Kampert JB, Kohl HW 3rd, Barlow CE, Macera CA, Paffenbarger RS Jr, Gibbons LW: Influences of cardiorespiratory fitness and other precursors on cardiovascular disease and all-cause mortality in men and women. JAMA 1996;276:205–210.

112 Barlow CE, LaMonte MJ, Fitzgerald SJ, Kampert JB, Perrin JL, Blair SN: Cardiorespiratory fitness is an independent predictor of hypertension incidence among initially normotensive healthy women. Am J Epidemiol 2006;163:142–150.
113 Blair SN, Kohl HW 3rd, Paffenbarger RS Jr, Clark DG, Cooper KH, Gibbons LW: Physical fitness and all-cause mortality: a prospective study of healthy men and women. JAMA 1989;262:2395–2401.
114 Stevens J, Cai J, Evenson KR, Thomas R: Fitness and fatness as predictors of mortality from all causes and from cardiovascular disease in men and women in the lipid research clinics study. Am J Epidemiol 2002;156:832–841.
115 McAuley P, Pittsley J, Myers J, Abella J, Froelicher VF: Fitness and fatness as mortality predictors in healthy older men: the veterans exercise testing study. J Gerontol A Biol Sci Med Sci 2009;64:695–699.
116 Speakman JR, Hambly C: Starving for life: what animal studies can and cannot tell us about the use of caloric restriction to prolong human lifespan. J Nutr 2007;137:1078–1086.
117 Gaesser GA: Exercise for prevention and treatment of cardiovascular disease, type 2 diabetes, and metabolic syndrome. Curr Diab Rep 2007;7:14–19.
118 Ingram DK, Roth GS, Lane MA, Ottinger MA, Zou S, de Cabo R, Mattison JA: The potential for dietary restriction to increase longevity in humans: extrapolation from monkey studies. Biogerontology 2006; 7:143–148.
119 Colman RJ, Anderson RM, Johnson SC, Kastman EK, Kosmatka KJ, Beasley TM, Allison DB, Cruzen C, Simmons HA, Kemnitz JW, Weindruch R: Caloric restriction delays disease onset and mortality in rhesus monkeys. Science 2009;325:201–204.
120 Fontana L, Meyer TE, Klein S, Holloszy JO: Long-term low-calorie low-protein vegan diet and endurance exercise are associated with low cardiometabolic risk. Rejuvenation Res 2007;10:225–234.
121 Willcox DC, Willcox BJ, Todoriki H, Curb JD, Suzuki M: Caloric restriction and human longevity: what can we learn from the Okinawans? Biogerontology 2006;7:173–177.

Derek M. Huffman, PhD
Institute for Aging Research, Department of Medicine, Division of Endocrinology
Golding Building, Room 502, The Albert Einstein College of Medicine
1300 Morris Park Avenue, Bronx, NY 10461 (USA)
Tel. +1 718 430 4278, Fax +1 718 430 8557, E-Mail derek.huffman@einstein.yu.edu

Mobbs CV, Hof PR (eds): Body Composition and Aging.
Interdiscipl Top Gerontol. Basel, Karger, 2010, vol 37, pp 175–192

Clinical, Cellular and Molecular Phenotypes of Aging Bone

Farhan A. Syed[a] · Jameel Iqbal[b] · Yuanzhen Peng[b] · Li Sun[b] · Mone Zaidi[b]

[a]College of Medicine, Mayo Clinic, Rochester, Minn., and [b]The Mount Sinai Bone Program, and Department of Medicine, Mount Sinai School of Medicine, New York, N.Y., USA

Abstract

Our understanding of gerontological bone loss and osteoporosis has grown substantially in the recent past. Clinical as well as basic and translational studies have been pivotal in providing us with the pathophysiology of this condition. They have also informed us of the various cellular and molecular mechanisms underlying age related bone loss. This chapter focuses on the current concepts and paradigms of age related bone loss in humans and how various animal and cellular models have broadened our understanding in this fascinating but complex area. Changes in hormonal, neuronal and biochemical cues with age and their effect on bone have been discussed. This chapter also outlines recent studies on the relationship between bone and fat in the marrow, and the fate of the marrow mesenchymal stromal cell population which can give rise to either bone forming osteoblasts or fat-forming adipocytic cells as a function of age.

Osteoporosis or porous bone disease is defined as a systemic skeletal disease characterized by low bone mass and microarchitectural deterioration of bone tissue, with a consequent increase in bone fragility and susceptibility to fracture. Traditionally, primary osteoporosis refers to both postmenopausal osteoporosis in women and senile osteoporosis in both elderly men and women, which have been referred to as type I and type II, respectively [1, 2]. Our understanding of the pathophysiology and gerontology of age-related bone loss has gained immensely from clinical observations and human studies, as well as basic and translational research involving both cellular and animal models.

Bone is a dynamic organ with both mechanical and metabolic functions and its composition reflects this role [3]. The bone composite comprises of organic and inorganic phases with its extracellular matrix consisting of hydroxyapatite mineral, collagenous and noncollagenous proteins, water and also varying amounts of lipid which depends on the age, species and site [3]. At the cellular level, bone is composed of three structurally and functionally distinct cell types, namely the bone-forming osteoblasts, bone-resorbing osteoclasts and bone-regulating osteocytes. In the adult skeleton, osteoblasts account for 4–6%, osteoclasts for 1–2% and osteocytes for more than 90% of the total bone cells [4]. Bone is being constantly turned over in distinct areas of the skeleton termed the bone metabolic units (BMUs) [5, 6]. In a healthy young adult skeleton, the processes of bone resorption and formation at a given BMU are tightly coupled such that the rate of bone formation and matrix mineralization equals the rate of matrix degradation and mineral dissolution, thereby maintaining a steady state of bone homeostasis [6]. However, during aging, significant amounts of bone are lost due to the tipping of this delicate balance towards enhanced resorption coupled with decreased formation leading to net bone loss in the aging population; this ultimately manifests as osteoporosis [5].

Bone loss in humans can be represented by quantitative (bone mineral density) loss, as well as qualitative detriments in the bone geometry and microstructure that occur in both sexes, albeit at different rates and time points. Intrinsic factors such as genetics and peak bone mass accrual in youth and alteration in hormonal and biochemical cues determine bone changes with age. In addition, a number of extrinsic challenges, such as nutrition and life style choices, play key roles. The presence of other secondary diseases or pathologies such as diabetes, hyperparathyroidism, inflammatory and auto immune diseases, such as rheumatoid arthritis (RA), also contribute to the extent, severity and timing of bone loss. Certain drugs to treat these conditions, such as thiazolidinediones (TZDs) for diabetes and glucocorticoids for RA negatively impact both the quantity and quality of bone.

In this chapter, we will review the current concepts of age-related bone loss in humans and the various rodent models, especially from genetically engineered mice, that have been utilized in studies on aging. We will also specifically outline recent studies which further our understanding of the cellular and molecular changes that occur in bone and bone marrow with age, and how these are regulated either systemically or peripherally by various hormonal, biochemical and neuronal cues.

Quantitative and Qualitative Changes in the Aging Bone

The adaptation of the skeleton to the demands of weight bearing and propulsion require rigidity and flexibility. Thus, the skeleton is made up of dense matrix impregnated with hydroxyapatite that determines the quantity of bone, as assessed clinically by measuring bone mineral density (BMD) at mainly cortical or cancellous sites. The quality of bone is determined by its shape, size, geometry and microarchitecture in its cancellous and cortical bone compartments [7]. During adult life bone is being constantly remodeled and redistributed to achieve strength commensurate with its loading requirements [8]. Detrimental changes in the skeleton with age in both men and women are well established [1, 2]. These changes can be either due to a decrease in the quantity, as assessed by decreases in bone mineral density (BMD) at almost all skeletal sites [9], or due to qualitative and structural changes in the skeleton with age [7]. In fact, the observation that the risk for fracture increases with age independently of BMD [10], has led to the recognition, for example, that changes in the microarchitecture of trabecular bone as well as those in the geometric properties of cortical bone are critical determinants of bone fragility [7]. In addition to decreases in trabecular and cortical BMD, the aging skeleton also displays cortical thinning, cortical porosity, impaired mineralization, cement line accumulation, and trabecular thinning, perforation and loss [11]. This results in a progressive accumulation of microcracks, which further damage the canilicular system leading to osteocyte apoptosis [12]. All these changes negatively impact bone homeostasis, and ultimately strength [7].

Genetics of Peak Bone Mass Accrual in Subsequent Bone Loss with Aging

Osteoporosis is a polygenic disease with each gene having small to moderate effect on various parameters and fracture risk [13]. Studies with monozygotic and dizygotic twins have pointed to the high heritability of various bone structural parameters (up to 80%) [13]. For example, close relatives of osteoporotic patients display reduced BMDs if not osteopenia compared with age- and gender-matched controls [14]. These observations have led to attempts at identifying genes linked to disease etiology using the candidate gene approach or linkage and association studies.

The vitamin D receptor gene (VDR) was one of the first to be identified using a candidate gene approach; however, the original findings were not replicable in other populations [15]. Similar polymorphisms were also described for ERα, Col1a1 and other bone-related proteins and receptors [13]. The genome-wide scan approach has implicated the Wnt signaling pathway [16], in addition to finding a strong association between the BMP-2 gene and a combined phenotype of low BMD and fracture in families of Icelandic ancestry [13]. Such studies point to the importance of genetics in rates of bone loss with age and susceptibility to fractures in humans.

What is also equally important is how genetics can impact bone gain in young adulthood and in the attainment of peak bone mass. Epidemiologic studies have demonstrated a relationship between birth weight, weight in infancy and adult bone mass [17]. Several longitudinal studies that follow bone mass accrued through childhood and adolescence and the use of mathematical models suggest that modifying peak bone mass in young adulthood will result in bone loss and fragility at an advanced age [17]. However, the definition and rate of accrual of peak bone mass and rate of age-related bone loss must take into account factors such as ethnicity, body size and shape [18].

Impact of Environmental Factors on Bone Loss with Age

Nutritional and lifestyle choices affect the skeleton both in early life as well as in utero [17]. Statistically significant relationships between low body weight probably caused by undernutrition in infancy, and bone mass have been reported from independent studies [19, 20]. Calcium intake is critical for maintaining bone mass in adulthood; low calcium intakes can lead to PTH-stimulated bone remodeling and increase in resorption. Calcium and vitamin D prevent hip fractures in elderly women [21] and men [22]. Other environmental and lifestyle factors, such as smoking, inactivity and loss of muscle mass also impact the rate and severity of bone loss with age. Likewise, medications such as glucocorticoids, TZDs, anticonvulsants and SSRIs have been shown to cause bone loss in the elderly population [23].

Patterns of Age-Related Bone Loss in Men and Women

Although men and women lose bone as a consequence of aging, the patterns of loss differ between the genders. Early menopause is associated with a dramatic bone loss, which is transient and may last up to a decade. It is primarily trabecular loss (20–30%), with cortical loss being less than half [25]. While this loss is established as being primarily due to declining estrogen levels [24], elevated follicle-stimulating hormone (FSH) levels, even before serum estrogen has dropped, are thought to directly enhance osteoclastic bone resorption and cause bone loss, in addition to causing elevations in pro-osteoclastogenic cytokines [25–30]. In fact, as early as the late perimenopause, there is evidence of accelerated osteoclastic bone resorption, which correlates with high serum FSH levels [31–33].

The Study of Women Across Nations (SWAN), a cross-sectional and longitudinal study of over 2,300 perimenopausal women between 42 and 52 years, showed that not only was there a strong correlation between basal FSH and bone remodeling markers [31], but a change in FSH levels over 4 years could predict a change in bone mass [34]. It is indeed difficult to establish causality from SWAN, as high FSH is a more sensitive marker for hypogonadism than reduced estrogen; hence, this association may simply be correlative.

However, a step towards establishing causality is a smaller study, which shows that, with similar estrogen levels, amenorrheic women having a mean serum FSH of 35 IU/l had considerably greater bone loss than women whose mean serum FSH level was 8 U/l [35]. Thus, it appears that FSH has a role in causing hypogonadal bone loss that has been traditionally attributed solely to low estrogen. It also deserves mention that menopause also affects estrogen-related signals including activin and inhibin that are involved in regulation of FSH production, for which receptors are also present on bone cells including osteoblasts and osteoblast precursors [36]. More recent evidence suggests that both follistatin and activin A regulate osteoblastic bone formation [37].

Men undergo a continuous, but slow phase of bone loss [38]. In fact, the accelerated bone loss in women, which decreases exponentially over a decade, merges asymptotically with the subsequent slow phase that continues indefinitely. The slow phase of bone loss is characterized by secondary hyperparathyroidism and rise in bone turnover markers that are directly correlated to the rise in PTH. It is also clear that, in men, biologically available sex steroid levels decrease substantially due primarily to an increase in binding proteins, such as sex hormone-binding globulin (SHBG). Only sex steroids that are either free (unbound) or bound to albumin are biologically active. Over the male lifespan, the bioavailable estrogen and testosterone decline by 47 and 64%, respectively [39]. Although testosterone is a major sex steroid in men, there is strong evidence from both cross-sectional [40] and longitudinal studies [41] that serum bioavailable estradiol levels rather than testosterone levels correlate better with preservation of BMD at various sites in elderly men. Finally, age-related bone loss in both genders is also associated with reductions in the production of hormones and growth factors important for osteoblast differentiation and function. For example, aging reduces the amplitude and frequency of growth hormone secretion [42], decreases serum IGF-1 and IGF-2 levels [43], and increases IGFBP-2 [44], all of which correlates inversely with bone mass in elderly subjects.

Murine Models of Normal Aging

An authentic model of senile osteoporosis can be argued to be normally aging mice or rats, which undergo bone loss as a function of age. It has been known for almost three decades that lipofuscin, an age pigment, accumulates in aging preosteoblasts, osteoblasts and osteocytes in mice of either gender [45–47]. Commonly used laboratory mice have a life span of 2–3 years, they attain peak bone mass between 4 and 8 months of age, and this is followed by a steady decline in bone mass as the mice age [48]. For example, C57BL/6 mice exhibit decreased cancellous and cortical bone mass and quality as a function of age [46, 47, 49]. There are subtle differences in the rate of cortical and cancellous bone loss in these aging mice, with the cancellous bone volume fraction decreasing from 6 weeks to 24 months, while cortical thickness increases paralleling the attainment of peak bone mass at around 6 months of age followed then by a progressive decline [46]. This age-related loss in cortical thickness is also observed in both genders with age accounting for 52.9 and 37.9% of the variance in females and males, respectively [50].

At the cellular level, there are changes in the osteoblastogenic and osteoclastogenic potential of the respective stem cells in the bone marrow. For example, 24-month-old BALB/c mice display decreases in the proliferative potential of osteogenic stem cells [51]. Likewise, aged C57BL/6 mice, in addition to decreases in the proliferation of their osteoblast precursors, also have a significant impairment in differentiation potential [52]. Such studies suggest that decreases in the osteogenic potential of bone marrow stromal stem cells cause the decreases in bone formation [51, 52]. It has also been shown that hematopoiesis is affected by aging [53]; thus, mice develop clonal B cell expansion [54] and lymphoma with advanced age [55]. It is interesting to note that the osteoclast differentiation factor RANKL, which is necessary for osteoclastogenesis and is secreted by B and T cells, is reported to increase with age, correlating with changes in cancellous bone volume [47]. However, not all studies agree and some actually show a decrease in serum RANKL with aging in humans [56] or rats [57]. Although it is generally recognized that aging is primarily associated with a decrease in osteoblasts, studies in aging mice have also reported increases in osteoclasts with age [57].

After 12 months of age, male and female Wistar rats display bone loss both at trabecular and cortical sites along with cortical thinning [58, 59]. Interestingly, the magnitude of bone lost depended on site, at least in females [59]. Recently, Duque et al. [60] described the bone phenotype in a rat model of healthy aging, the LOU/c rat, which displays a relatively increased lifespan in the absence of obesity and a low incidence of age-related diseases. The authors reported that aging resulted in reduced osteoblast and osteoclast numbers along with a significant increase in the marrow adiposity of long bones. These changes at the tissue and cellular levels correlated with decreases in serum C-telopeptide and serum osteocalcin, markers used to assess bone resorption and formation, respectively [60]. The model offers promise for future studies on the mechanisms of senile osteoporosis in an otherwise healthy population.

It has to be noted, nonetheless, that species- and strain-specific effects do remain. For example, decreases in bone formation seem to be a commonly observed phenomenon in both these experimental species [51, 52, 60]; however, a decrease in osteoclast number and function is noted in rats [60], but not in mice [47, 57]. Furthermore, different inbred mouse strains display a range of peak bone phenotypes that are genetically determined and are compartment-specific [61]; this can potentially impact the rate at which absolute amount of bone is lost with age.

Gonadectomized Models of Bone Loss

Gonadectomy in either sex is associated with an increase in bone remodeling, increased bone resorption, and a relative deficit in bone formation, resulting in accelerated bone loss [62]. It can be argued that these models cannot be classified as representing senile osteoporosis, but rather mimic postmenopausal bone loss in women or castration in men. Nonetheless, they have been routinely used to understand mechanisms of bone loss with age as sex steroid loss is an important contributing factor to bone loss in the aging population [2]. In fact, in mice and rodents, as in humans, the skeletal response to sex steroids is dose dependent. Low doses of estrogen prevent bone loss by reducing bone resorption, while higher doses stimulate bone formation [63]. It has also been suggested from human studies [64] and from estrogen receptor knockout mice that estrogens may be more protective than androgens [65] in both genders in preventing age-related bone loss. Estrogen or androgen loss can accelerate the effects of aging on bone by decreasing the antioxidant defense mechanisms in aging C57BL/6 mice, thereby resulting in increased oxidative stress [66]. Estrogen can also modulate the expression of other nuclear hormone receptors, such as the vitamin D receptor (VDR): ovariectomized old mice have reductions in the expression of the VDR in osteoblasts, and estrogen supplementation in these mice reverses this phenotype more so in the old than in young mice [67]. Of note is that ovariectomy in mice or rats leads to decreases in bone volume and increases in adipose volume in the bone marrow, a phenotype strikingly similar to that seen in aged humans [68, 69]. It has also been established using paired iliac crest biopsies that marrow adipocyte volume can significantly increase in elderly osteoporotic postmenopausal women and estrogen prevents this increase [70]. Such studies make a compelling argument to utilize ovariectomized rodent models as a surrogate for an aging bone phenotype, as long as the limitations are clearly understood and the complex interplay of sex hormone and glycoprotein signaling is appreciated.

The SAMP6 Mouse: A Model of Impaired Osteoblastogenesis and Bone Formation

The classic model of senescence acceleration and age-associated disorders is the senescence accelerated mouse or the SAM mouse [71]. A total of nine strains of accelerated aging were identified of which the SAMP6 strain has been established as a model for senile osteoporosis. It is characterized by low peak bone mass at maturation [72], which is polygenetically determined, a situation akin to osteoporosis in humans [63]. It was initially shown that SAMP6 mice display reduced cancellous bone formation rates due to a paucity of osteoblast progenitor cells in the marrow and that this phenotype becomes severe as these mice age [73]. Recently, however, it has been established that the impaired marrow osteogenesis is associated with reduced cortical bone formation at the endocortical, but not at the periosteal surface in long bones [74], and that there is a consistently low bone density across axial and appendicular bone sites [75].

An increase in resorption due to enhanced osteoclast maturation has also been reported [76]. However, it appears that this model is primarily one of impaired bone formation, as is suggested further by studies showing increased adipogenesis and myelopoiesis, and defective osteoblastogenesis [77]. Evidence that the defect is primarily in the osteoblast is indeed compelling: bone marrow stromal cells derived from SAMP6 mice display reduced binding and activity of the AP-1 transcription factor Jun-D. This results in lowered IL-11 expression, which likely impairs bone formation [78]. The Wnt pathway is also suppressed, and treatment with lithium chloride, a stimulator of the canonical Wnt pathway, enhances bone formation in vivo [79]. Conversely, the expression of secreted frizzled-related protein 4 (SFRP-4), an inhibitor of the Wnt pathway is enhanced in SAMP6 cells, thereby negatively regulating bone formation by inhibiting osteoblast proliferation and their responsiveness to ligands of the Wnt pathway such as Wnt 3A [80].

The Klotho-Deficient Mouse: A Model of Disrupted Osteoblasts, Osteoclasts, and Osteocytes and Defects in Mineral Metabolism

The *klotho* gene deficient mouse is also considered a valid model for an accelerated aging state mimicking a number of gerontological pathologies such as osteoporosis, skin atrophy, ectopic calcification and gonadal dysplasia [81]. The *klotho* gene encodes a single pass transmembrane protein that forms a complex with multiple fibroblast growth factor (FGF) receptors and is a co-receptor for FGF23, a bone-derived hormone that induces a negative phosphate balance [82]. Klotho deficiency causes irregular distributions of osteocytes and bone matrix proteins, as well as an accelerated aging of bone cells [83]. *Klotho–/–* mice display organelle-defective flat osteoblasts and small TRAP-positive osteoclasts at metaphyseal trabeculae in long bones [84]. A thick mineralized matrix in the proliferative zone of the growth plates and large nonmineralized areas in the metaphyseal trabeculae are present, suggesting that both cellular and mineralization activities are severely impaired [84]. These findings are consistent with an earlier report of impaired osteoblastic and osteoclastic differentiation leading to low-turnover osteopenia in these mice [85].

Klotho functions as an aging-suppressor and has been reported to be a regulator of oxidative stress and senescence [86]; in fact, loss of klotho leads to enhanced apoptosis in odontoblasts and disturbed mineralization in teeth [87]. In contrast to what had been described originally [81], it has been shown more recently by micro-CT that, in 3-month-old *klotho–/–* mice, trabecular bone mass at the tibia actually increases over wild type littermates [88]. Thus, there could be regional differences in the response to loss of klotho in the skeleton which may well involve its ability to modulate pathways, such as the Wnts and others involved in calcium and vitamin D homeostasis [88]. This model certainly warrants further study to explore mechanisms of age-related osteoporosis.

Mouse Models of Laminopathies and Extracellular Matrix Protein Deficiencies

Mutations in the *LMNA* gene which encodes for lamin A, a component of the nuclear lamina, leads to at least a dozen distinct human diseases collectively termed as laminopathies. These primarily arise from defects in the nuclear architecture [89]. Mice carrying an autosomal-recessive mutation in the lamin A gene display defects and a premature aging phenotype similar to Werner's and Hutchinson-Gilford Progeria syndromes [90]. The mice display trabecular and cortical bone defects in the axial and appendicular skeletons, consistent with an osteoporotic phenotype which is also observed in patients [90].

To elucidate the mechanisms of osteoporosis upon loss of lamins, laminin expression was reduced in normal human osteoblasts and in differentiating mesenchymal stem cells [91]. This was associated with lower osteoblast differentiation in marrow stromal cells and a decrease in the activity of mature osteoblasts. These osteoblastic defects were coupled with increased adipocyte differentiation in marrow stromal cells, which also displayed reduced Runx2 nuclear binding activity [91]. In agreement with these in vitro findings, mice knocked out for zmpste-24, a zinc metalloproteinase involved in the post-translational processing of prelamin A to mature lamin A, also displayed a progeroid bone phenotype with reduced osteoblast and osteocyte numbers and a parallel increase in marrow adipocytes [92]. Given the fact that oxidative stress increases with age and lamins are susceptible to free radical modifications (such as protein glycations) leading to genomic instability [93], it can be argued that the loss of nuclear integrity and resultant modulation of transcription factor binding activity may play a key role in the etiology of age-related bone loss due to reduced osteoblast formation.

Extracellular matrix proteins play important roles in the regulation and signaling in bone cells [94]. Loss of bone short collagen IX in male and female mice leads to significant loss of trabecular bone in an age-dependent manner [95]. This progressive increase in resorption is accompanied by an increased RANKL/OPG ratio in osteoblasts and in enlarged and active osteoclasts [95]. On the other hand, loss of osteopontin in mice leads to increases in bone formation in vivo and mineralized nodule formation in vitro upon treatment with a prostaglandin E receptor agonist [96]. It is interesting that osteopontin expression in osteoblasts decreases with age, at least in rats, suggesting the potential to design anabolic drugs if a similar decrease is found in aging humans [97]. Additionally, biglycan-deficient mice develop age-dependent osteopenia and progressive osteoporosis [98]; this is related to decreases in the proliferation and commitment to osteoblasts coupled with an increased rate of apoptosis in progenitors [99]. Changes in cellular matrix elasticity and integrity can also be a potential mechanism of bone loss. For example, it has been shown that the loss of RhoA signaling and cytoskeletal integrity in bone marrow stromal cells in vitro [100] can lead to decreased osteogenesis and increased adipogenesis, a scenario similar to that observed in the aging marrow [70, 77].

Mice Genetically Modified for Peptides, Hormones and Transcription Factors: Implications for the Aging Skeleton

Genetically modified mice for hormones, peptides and their cognate receptors and downstream transcription factors have provided information on the aging skeleton. Estrogen receptors seem to affect the aging skeleton, as indicated by the enhanced trabecular bone volume and increased expression of the osteogenic transcription factor Runx2 in estrogen receptor-β knockout mice (ERβ–/–) [101]. Aged Runx2 haploinsufficient mice do not respond to high-dose estrogen [102]. Aged mice also require a full gene dosage of this transcription factor to recover cancellous bone volume after bone marrow ablation [103].

Transgenic mice overexpressing a dominant negative form of Runx2 (Δcbfa1) postnatally are also osteopenic due to reductions in bone formation rate due to lowered osteoblastic activity, rather than osteoblast number, suggesting that once osteoblasts are differentiated, cbfa1 can regulate their function [104]. Conversely, mice haploinsufficient for the adipogenic transcription factor PPARγ display increased bone mass associated with increased osteoblastogenesis and decreased adipogenesis [105]. Attenuation of bone morphogenetic protein pathways (BMPs) are also implicated in aging. For example, mice overexpressing the bone morphogenetic protein inhibitor noggin in mature osteoblasts show dramatic decreases in BMD and bone formation rates [106]. Deficiencies in the growth hormone-IGF-1 axis with age, accompanied by reduced levels of both growth hormone and IGF-1 can also contribute to senile osteoporosis [107].

Neural Regulation of the Aging Skeleton

There is growing evidence that bone remodeling is under regulation of the central nervous system (CNS) [108]. That leptin is a negative regulator of bone formation in vivo was established by the demonstration that *ob/ob* and *db/db* mice display a high bone mass phenotype with increased rates of bone formation, in the face of hypogonadism [109]. Likewise, intracerebroventricular infusion of low doses of leptin decreased trabecular bone mass in *ob/ob* and wild-type mice [108]. Since these original studies, a number of central neurotransmitters and receptors have been implicated in the control of bone remodeling, such as cocaine and amphetamine regulated transcript (CART), β-adrenergic receptors, and cannabinoid receptors [110–112]. Young CB1–/– mice have a high bone mass phenotype, primarily due to reduced bone resorption. However, on aging; marrow stromal cells from these mice display enhanced adipogenesis at the expense of osteogenesis by upregulating adipogenic signals, such as cAMP and pCREB [112].

However, not all studies with the central regulation have yielded similar results. A drop in serum leptin levels in aging mice is associated with decreases in bone mass and strength, and increases in serum markers of bone resorption [113]. Loss of peripheral leptin leads to decrements in bone mineral density, cortical thickness and trabecular volume in femurs with a parallel rise in adipocyte numbers. Surprisingly, at the lumbar vertebrae, increased trabecular bone parameters and decreased adipocyte numbers are observed pointing to site-specificity in leptin effects [114]. A recent study utilizing β-adrenergic receptor knockout mice suggested direct positive effects of weight and/or leptin on bone turnover and cortical bone, independent of adrenergic signaling [111].

Cellular and Molecular Changes in Bone Marrow with Aging

Osteoclasts are derived from hematopoietic stem cell precursors of the monocyte/macrophage lineage [115], while osteoblasts arise from the stromal cell compartment [116, 117]. Marrow stromal cells are multipotent cells that can potentially give rise not only to osteoblasts, but also to adipocytes, chondrocytes and myoblasts [118]. The exact sequence of progression from a state of stemness to a more advanced progenitor is not clear [119]. However, studies indicate that osteoblastic and adipocytic differentiation are closely linked and that a bipotential osteoblast/adipocyte precursor exists whose commitment to either lineage depends on the strength of signals that favor either lineage [120]. For example, the transcription factors Runx2 and osterix promote osteoblastic [104, 121], while C/EBPβ and PPARγ promote adipocytic commitment and differentiation [122]. That the balance between such lineage-specific nuclear transcription factors is important was highlighted by findings in mice with haplo-insufficiency of PPARγ; these mice have an increase in bone mass associated with increased osteoblastogenesis and decreased adipogenesis [105]. Collectively, these findings have led to the concept of plasticity between osteoblast and adipocyte lineages, and it is becoming increasingly clear that a change in bone marrow stromal cell dynamics can result in osteoporosis with age due to an increase in the number of marrow adipocytes at the expense of osteoblasts [123].

The differentiation of the osteoblast/adipocyte bipotential stromal cell into either the osteoblastic or adipocytic phenotype depends on the signals which it receives, including hormonal and cytokine stimuli, among others, in the marrow microenvironment [120, 124, 125]. For example, various BMPs [125] and Wnt signals [126] induce osteoblast differentiation both independently and synergistically, while Dlk-1/Pref-1 [127] and noggin [128] inhibit the same process. Likewise, marrow adipogenesis may be stimulated or inhibited depending upon the cues present in the stromal cell milieu. BMP4 and insulin are known inducers of adipocytic differentiation both in the periphery and in marrow [129, 130]. So also are natural ligands to PPARγ, such as prostaglandin J2 derivatives [131]. It is interesting that murine osteoblasts derived from aging bone marrow display significant reduction in Wnt pathway genes, and an upregulation of the Wnt antagonist DKK-1 [132]. Coupled with this is the downregulation of the pro-osteogenic TGF-β pathway [133]. Increased lipid oxidation with aging has been shown to cause oxidative stress, increased expression of PPARγ and diminished pro-osteogenic Wnt signaling in the skeleton. The latter likely involves FoxO transcription factors, which play important roles in maintaining the antioxidant status in aging cells [134]. In fact, PPARγ2 controls multiple regulatory pathways of osteoblast differentiation from marrow stromal cells and is thus being proposed as a master regulator of osteogenesis/adipogenesis in the marrow [135].

As regards the role of sex steroids and peptide hormones, estrogen promotes osteoblastogenesis and inhibits adipocytic differentiation of bone marrow derived stromal cell lines [136]. Loss of estrogens in aged osteoporotic postmenopausal women is associated with a significant increase in bone marrow adipocyte volume and treatment with an estrogen patch for one year results in significant decreases in marrow adipocyte volume and prevention of adipocyte hyperplasia [70]. In the same cohort of patients, estrogen treatment increased bone mineral density [117]. Human mesenchymal stromal cells also express FSH receptors [25, 137]; it is conceivable that FSH may directly affect bone formation.

The exact aging microenvironment seen by bone marrow stromal cells is not yet well defined. Attempts have been made to understand this in vitro by using conditioned media [119]. Exposing marrow stromal cells in vitro to serum obtained from young and old donors [138] showed that sera of elderly donors was inhibitory to osteoblast and permissive to adipocyte differentiation; however, the putative factor(s) responsible for these effects need to be determined [96].

Post et al. [139] have challenged the reciprocity of the osteoblast/adipocyte switching. Using two plastic-adherent clonal cell lines mMSC1 and mMSC2 derived from murine bone marrow, they show that, while the clone mMSC1 differentiates to adipocytes, the cells cannot be induced to differentiate to osteoblasts when treated with osteogenic media. In contrast, while the mMSC2 clone could differentiate into osteoblasts and form bone in vivo in immunodeficient mice, it could not be induced to differentiate into an adipocytic lineage, even under stringent adipocytic stimuli. This has led to the suggestion that distinct clonal subpopulations of cells committed to either osteoblast or adipocyte lineage exist in the marrow and can undergo independent changes during aging. The way to reconcile what is known so far and the observations of Post et al. [139] is that the presence of these two distinct subpopulations of marrow pre-osteoblastic/pre-adipocytic clones does not imply the absence of a third truly bipotential osteoblastic/adipocytic population, amenable to differentiation either way depending on whether the cues are pro-osteogenic or pro-adipogenic. This remains a fascinating and therapeutically important phenomenon that is likely to be an active area of investigation in the future.

Conclusions

Age-related osteoporosis is a multi-faceted disease, which involves a number of distinct tissue pathologies, such as decreases in trabecular and cortical bone density, volume and thickness, and a disorganization of the extracellular matrix. There are a multitude of studies in humans, genetically modified mice, and bone cells highlighting the multiple molecular pathways that undergo an age-related change. These are by no means exhaustive or complete. However, what is becoming clear is that although the aging bone phenotype can be impacted by gender, race and genetics, the underlying cellular phenotype seems to be consistent. As a function of age in both mice and man, we consistently observe: decreases in bone volume and increases in marrow fat, degeneration of marrow components and poor extracellular matrix quality, and an upregulation of pro-adipogenic and anti-osteogenic signals which switch stromal cells towards adipocytes and away from osteoblasts. These animal and cellular models of bone aging enable the recognition of future targets for drug design.

References

1 Riggs BL, Wahner HW, Seeman E, Offord KP, Dunn WL, Mazess RB, Johnson KA, Melton LJ III: Changes in bone mineral density of the proximal femur and spine with aging: differences between the postmenopausal and senile osteoporosis syndromes. J Clin Invest 1982;70:716–723.

2 Riggs BL, Khosla S, Melton LJ III: The type I/type II model for involutional osteoporosis: update and modifications based on new observations; in Marcus R, Feldman D, Kelsey J (eds): Osteoporosis, ed 2. New York, Academic Press, 2001, pp 49–58.

3 Robey PG, Boskey AL: The composition of bone; in: Primer on the Metabolic Bone Diseases and Disorders of Mineral Metabolism, ed 7. Washington, American Society for Bone and Mineral Research, 2008, pp 32–38.

4 Bonewald LF: Osteocytes; in: Primer on the Metabolic Bone Diseases and Disorders of Mineral Metabolism, ed 7. Washington, American Society for Bone and Mineral Research, 2008, pp 32–38.

5 Riggs BL, Khosla S, Melton LJ III: Sex steroids and the construction and conservation of the adult skeleton. Endocr Rev 2002;23:279–302.

6 Lee CA, Einhorn TA: The Bone Organ System: Form and Function; in Marcus R, Feldman D, Kelsey J (eds): Osteoporosis, ed 2. New York, Academic Press, 2001, pp 2–20.

7 Seeman E: Structural basis of growth-related gain and age related loss of bone strength. Rheumatology 2008;47:iv2–iv-8.

8 Szulc P, Seeman E: Thinking inside and outside the envelopes of bone. Osteoporos Int 2009;20:1281–1288.

9 Melton LJ III, Ilstrup DM, Beckenbaugh RD, Riggs BL: Hip fracture recurrence: a population-based study. Clin Orthop 1982;167:131–138.

10 Cummings SR, Nevitt MC, Browner WS, Stone K, Fox KM, Ensrud KE, Cauley J, Black D, Vogt TM: Risk factors for hip fracture in white women: study of Osteoporotic Fractures Research Group. N Engl J Med 1995;332:767–773.

11 Marcus R, Majumdar S: The nature of osteoporosis; in Marcus R, Feldman D, Kelsey J (eds): Osteoporosis, ed 2. New York, Academic Press, 2001, pp 49–58.

12 Hazenberg JG, Freeley M, Foran E, Lee TC, Taylor D: Microdamage: a cell transducing mechanism based on ruptured osteocyte processes. J Biomech 2006;39:2096–2103.

13 Center JR, Eisman JA: Genetics of osteoporosis; in: Primer on the Metabolic Bone Diseases and Disorders of Mineral Metabolism, ed 7. Washington, American Society for Bone and Mineral Research, 2008, pp 32–38.

14 Evans RA, Marel GM, Lancaster EK, Kos S, Evans M, Wong SY: Bone mass is low in relatives of osteoporotic patients. Ann Intern Med 1988;109:870–873.

15 Morrison NA, Qi JC, Tokita A, Kelly PJ, Crofts L, Nguyen TV, Sambrook PN, Eisman JA: Prediction of bone density from vitamin D receptor alleles. Nature 1994;367:284–287.

16 Gong Y, Slee RB, Fukui N, et al: LDL-receptor related protein 5 (LRP5) affects bone accrual and eye development. Cell 2001;107:513–523.

17 Cooper C, Westlake S, Harvey N, Javaid K, Dennison E, Hanson M: Review: developmental origins of osteoporotic fracture. Osteoporos Int 2006;17:337–347.

18 Schonau E: The peak bone mass concept: is it still relevant? Pediatr Nephrol 2004;19:825–831.

19 Dennison EM, Syddall HE, Sayer AA, Gilbody HJ, Cooper C: Birth weight and weight at 1 year are independent determinants of bone mass in the seventh decade: the Hertfordshire cohort study. Pediatr Res 2005;57:582–586.

20 Antoniades L, MacGregor AJ, Andrew T, Spector TD: Association of birth weight with osteoporosis and osteoarthritis in adult twins. Rheumatology 2003;42:791–796.

21 Chapuy MC, Arlot ME, Duboeuf F, Brun J, Crouzet B, Arnoud S, Delmas PD, Meunier PJ: Vitamin D_3 and calcium to prevent hip fractures in elderly women. N Engl J Med 1992;327:1637–1642.
22 Dawson-Hughes B, Harris SS, Krall EA, Dallal GE: Effect of calcium and vitamin D supplementation on bone density in men and women 65 years of age or older. N Engl J Med 1997;337:670–676.
23 Kiel DP, Rosen CJ, Dempster D: Age related bone loss; in: Primer on the Metabolic Bone Diseases and Disorders of Mineral Metabolism, ed 7. Washington, American Society for Bone and Mineral Research, 2008, pp 32–38.
24 Khosla S, Atkinson EJ, Melton LJ III, Riggs BL: Effects of age and estrogen status on serum parathyroid hormone levels and biochemical markers of bone turnover in women: a population based study. J Clin Endocrinol Metab 1997;82:1522–1527.
25 Sun L, Peng Y, Sharrow AC, et al: FSH directly regulates bone mass. Cell 2006;125:247–260.
26 Sowers MR, Greendale GA, Bondarenko J, et al: Endogenous hormones and bone turnover markers in pre- and perimenopausal women: SWAN. Ost Internat 2003;14:191–197.
27 Iqbal J, Sun L, Kumar TR, et al: Follicle stimulating hormone stimulates TNF production from immune cells to enhance osteoblast and osteoclast formation. Proc Natl Acad Sci USA 2006;103:14925–14930.
28 Zaidi M, Blair H C, Iqbal J, Kumar TR, Zallone A, Sun L: Proresorptive actions of FSH and bone loss. Ann NY Acad Sci 2007;1116:376–382.
29 Cannon JG, Cortez-Cooper M, Meaders E, Stallings J, Haddow S, Kraj B, Sloan G, Mulloy A: Follicle stimulating hormone, interleukin-1 and bone density in adult women. Am J Physiol Regul Integr Comp Physiol 2010;298:R790–R798.
30 Zaidi M, Turner CH, Canalis E, Pacifici R, Sun L, Iqbal J, Guo XE, Silverman S, Epstein E, Rosen CJ: Bone loss or lost bone: rationale and recommendations for the diagnosis and treatment of early postmenopausal bone loss. Curr Osteoporosis Rep 2009; 7:118–126.
31 Wu Y, Torcheria J, Yao W, Lane NE, Lanier LL, Nakamura MC, Humphrey MB: Bone microenvironment specific roles of ITAM adapter signaling during bone remodeling induced by acute estrogen deficiency. PLoS Ibe 2007;2:e586.
32 Wu XY, Wu XP, Xie H, Zhang H, Peng YQ, Yuan LQ, Sum X, Luo XH, Liao EY: Age-related changes in biochemical markers of bone turnover and gonadotropins levels and their relationship among Chinese adult women. Osteoporosis Int 2010;21:275–285.
33 Akhter MP, Lappe JM, Davies KM, Recker RR: Transmenopausal changes in trabecular bone structure. Bone 2007;41:111–116.
34 Sowers MR, Jannausch M, McConnell D, et al: Hormone predictors of bone mineral density changes during the menopausal transition. J Clin Endocrinol Metabol 2006;91:1261–1267.
35 Devleta B, Adem B, Seneda S: Hypergonadotropic amenorrhea and bone density: new approach to an old problem. J Bone Min Res 2004;22:360–364.
36 Gaddy-Kurten D, Coker JK, Abe E, et al: Inhibiin suppresses and activin stimulates osteoblastogenesis and osteoclastogenesis in murine bone marrow cultures. Endocrinology 2002;143:74–83.
37 Eijken M, Swagemakers S, Koedam M, et al: Activin A-follistatin system: potent regulator of human extracellular matrix mineralization. FASEB J 2007; 21:2949–2960.
38 Riggs BL, Khosla S, Melton LJ III: A unitary model for involution osteoporosis:estrogen deficiency causes both type I and type II osteoporosis in postmenopausal women and contributes to bone loss in aging men. J Bone Miner Res 1998;13:763–773.
39 Khosla S, Melton LJ III, Atkinson EJ, OFallon WM, Klee GG, Riggs BL: Relationships of serum sex steroid levels and bone turnover markers with bone mineral density in men and women: a key role for bioavailable estrogen. J Clin Endocrinol Metab 1998;83:2266–2274.
40 Szulc P, Munoz F, Claustrat B, garnero P, Marchand F: Bioavailable estradiol may be an important determinant of osteoporosis in men: the MINOS study. J Clin Endocrinol Metab 2001;86:192–199.
41 Khosla S, Melton LJ III, Atkinson, OFallon WM: Relationship of serum sex steroid levels to longitudinal changes in bone density in young versus elderly men. J Clin Endocrinol Metab 2001;86:3555–3561.
42 Thorner MO, Vance ML: Growth Hormone. J Clin Invest 1988;82:745–747.
43 Bennett J, Wahner HW, Riggs BL, Hintz RL: Insulin like growth factors I and II: aging and bone density in women. J Clin Endocrinol Metab 1984;59:701–704.
44 Amin S, Riggs BL, Atkinson EJ, Oberg AL, Melton LJ III, Khosla S: A potentially deleterious role of IGFBP-2 on bone density in aging men and women. J Bone Miner Res 2004;19:1075–1083.
45 Tonna EA: Accumulation of lipofuscin (age pigment) in aging skeletal connective tissues as revealed by electron microscopy. J Gerontol 1975;30:3–8.
46 Bikle DD, Sakata T, Leary C, Elalieh H, Ginzinger D, Rosen CJ, Beamer W, Majumdar S, Halloran BP: Insulin-like growth factor 1 is required for the anabolic actions of parathyroid hormone on mouse bone. J Bone Miner Res 2002;17:1570–1578.

47 Cao J, Venton L, Sakata T, Halloran BP: Expression of RANKL and OPG correlates with age-related bone loss in male C57BL/6 mice. J Bone Miner Res 2003;18:270–277.
48 Watanabe K, Hishiya A: Mouse models of senile osteoporosis. Mol Aspects Med 2005;26:221–231.
49 Ferguson VL, Ayers RA, Bateman TA, Simske SJ: Bone development and age related bone loss in male C57BL/6 mice. Bone 2003;33:387–398.
50 Cooper DML, Thomas CDL, Clement JG, Turinsky AL, Sensen CW, Hallgrimsson B: Age dependent change in the 3D structure of cortical porosity at the human femoral midshaft. Bone 2007;40:957–965.
51 Bergman RJ, Gazit D, Kahn AJ, Gruber H, McDougall S, Hahn TJ: Age related changes in the osteogenic stem cells in mice. J Bone Miner Res 1996;11:568–577.
52 Zhang W, Ou G, Hamrick M, Borke J, Wenger K, Chutkan N, Yu J, Mi QS, Isales CM, Shi XM: Age-related changes in the osteogenic differentiation potential of mouse bone marrow stromal cells. J Bone Miner Res 2008;23:1118–1128.
53 Morrison SJ, Wandycz AM, Akashi K, Globerson A, Weissman IL: The aging of hematopoietic stem cells. Nat Med 1996;2:1011–1016.
54 LeMaoult J, Manavalan JS, Dyall R, Szabo P, Nikolic-Zugic J, Weksler ME: Cellular basis of B cell clonal populations in old mice. J Immunol 1999;162:6384–6391.
55 Ghia P, Melchers F, Rolink AG: Age dependent changes in B lymphocyte development in man and mouse. Exp Gerontol 2000;35:159–165.
56 Kerschan-Schindl K, Wendlova J, Kudlacek S, Gleiss A, Woloszczuk W, Pietschmann P: Serum levels of receptor activator of nuclear factor kappaB ligand (RANKL) in healthy women and men. Exp Clin Endocrino Diabetes 2008;116:491–495.
57 Pietschmann P, Rauner M, Sipos W, Kerschan-Schindl K: Osteoporosis: An age related and gender specific disease: a mini review. Gerontology 2009;55: 3–12.
58 Lida H, Fukuda S: Age related changes in bone mineral density, cross sectional area and strength at different skeletal sites in male rats. J Vet Med Sci 2002; 64:29–34.
59 Fukuda S, Lida H: Age related changes in bone mineral density, cross sectional area and strength of long bones in the hind limbs and first lumbar vertebra in female Wistar rats. J Vet Med Sci 2004;66:755–756.
60 Duque G, Rivas D, Lei W, Li A, Henderson JE, Ferland G, Gaudreau P: Age-related bone loss in the LOU/c rat model of healthy aging. Exp Gerontol 2009;44:183–189.
61 Sabsovich j, Clark JD, Liao G, Peltz G, Lindsey DP, Jacobs CR, Yao W, Guo T-Z, Kingery WS: Bone microstructure and its genetic variability in 12 inbred strains: μCT study and in silico genome scan. Bone 2008;42:439–451.
62 Syed F, Khosla S: Mechanisms of sex steroid effects on bone. Biochem Biophys Res Commun 2005;328: 688–696.
63 Priemel M, Schilling AF, Haberland M, Pogoda P, Rueger JM, Amling M: Osteopenic mice: animal models of the aging skeleton. J Musculoskel Neuron Interact 2002;2:212–218.
64 Falahati-Nini J, Riggs BL, Atkinson EJ, OFallon WM, Eastell R, Khosla S: Relative contributions of testosterone and estrogen in regulating bone resorption and formation in normal elderly men. J Clin Invest 2000;106:1553–1560.
65 McCauley LK, Tozum TF, Rosol TJ: Estrogen receptors in skeletal metabolism: Lessons from genetically modified models of receptor function. Crit Rev Eukaryot Gene Expression 2002;12:89–100.
66 Almeida M, Han L, Martin-Millan M, Plotkin LI, Stewart SA, Roberson PK, Kousteni S, OBrien CA, Bellido T, Parfitt AM, Weinstein RS, Jilka RL, Manolagas SC: Skeletal involution by age-associated oxidative stress and its acceleration by loss of sex steroids. J Biol Chem 2007;282:27285–27297.
67 Duque G, El Abdaimi K, Macoritto M, Miller MM, Kremer R: Estrogens (E2) regulate expression and response of 1,25-dihydroxyvitamin D_3 receptors in bone cells: changes with aging and hormone deprivation. Biochem Biophys Res Commun 2002;299: 446–454.
68 Dunnill MS, Anderson JA, Whitehead R: Quantitative histological studies on age changes in bone. J Path Bact 1967;94:275–291.
69 Meunier P, Aaron J, Edouard C, Vignon A: Osteoporosis and the replacement of cell populations of the marrow by adipose tissue. Clin Orthop Rel Res 1971;80:147–154.
70 Syed FA, Oursler MJ, Hefferan TE, Peterson JM, Riggs BL, Khosla S: Effects of estrogen therapy on bone marrow adipocytes in postmenopausal osteoporotic women. Osteopor Int 2008;19:1323–1330.
71 Takeda T, Hosokawa M, Takeshita S, Irino M, Higuchi K, Matsushita T, Tomita Y, Yasuhira K, Hamamoto H, Shimizu K, Ishii M, Yamamuro T: New murine model of accelerated senescence. Mech Aging Dev 1981;17:183–194.
72 Matsushita M, Tsuboyama T, Kasai R, Okumura H, Yamamuro T, Higuchi K, Kohno A, Yunezo T, Utani A, Umezawa M, Takeda T: Age related changes in bone mass in the senescence accelerated mouse (SAM) SAM-R/3 and SAM/6 as new models for osteoporosis. Am J Pathol 1986;215:276–283.

73 Jilka RL, Weinstein RS, Takahashi K, Parfitt AM, Manolagas SC: Linkage of decreased bone mass with impaired osteoblastogenesis in a murine model of accelerated senescence. J Clin Invest 1996;97: 1732–1740.

74 Silva MJ, Brodt MD, Ko M, Abu-Amer Y: Impaired marrow osteogenesis is associated with reduced endocortical bone formation but does not impair periosteal bone formation in long bones of SAMP6 mice. J Bone Miner Res 2005;20:419–427.

75 Kasai S, Shimizu M, Matsumura T, Okudaira S, Matsushita M, Tsuboyama T, Nakamura T, Hosokawa M: Consistency of low bone density across bone sites in SAMP6 laboratory mice. J Bone Miner Metab 2004;22:207–214.

76 Okamoto Y, Takahashi K, Toriyama K, Takeda N, Kitagawa K, Hosokawa M, Takeda T: Femoral peak bone mass and osteoclast number in an animal model of age-related spontaneous osteopenia. Anat Rec 1995;242:21–28.

77 Kajnekova J, Lecka-Czernik B, Gubrij I, Hauser SP, Takahashi K, Parfitt AM, Jilka RL, Manolagas SC, Lipschitz DA: Increased adipogenesis and myelopoiesis in the bone marrow of SAMP6, a murine model of defective osteoblastogenesis and low turnover osteopenia. J Bone Miner Res 1997;12:1772–1779.

78 Tohjima E, Inoue D, Yamamoto N, Kido S, Ito Y, Kato S, Takeuchi Y, Fukumoto S, Matsumoto T: Decreased AP-1 activity and interleukin-11 expression by bone marrow stromal cells may be associated with impaired bone formation in aged mice. J Bone Miner Res 2003;18:1461–1470.

79 Clement-Lacroix P, Ai M, Morvan F, Roman-Roman S, Vayssiere B, Belleville C, Estrera K, Warman ML, Baron R, Rawadi G: Lrp-5 independent activation of Wnt signaling by lithium chloride increases bone formation and bone mass in mice. Proc Natl Acad Sci USA 2005;102:17406–17411.

80 Nakanishi R, Shimizu M, Mori M, Akiyama H, Okudaira S, Otsuki B, Hashimoto M, Higuchi K, Hosokawa M, Tsuboyama T, Nakamura T: Secreted frizzled-related protein 4 is a negative regulator of peak BMD in SAMP6 mice. J Bone Miner Res 2006;21:1713–1721.

81 Kuro-o M, Matsumura Y, Aizawa H, Kawaguchi T, Suga T, Utsugi Y, Ohyama M, Kurabayashi T, Kaname E, Kume E, Iwasaki H, Lida A, Shiraki-Lida T, Nishikawa S, Nagai R, Nabeshima YI: Mutation of the mouse klotho gene leads to a syndrome resembling aging. Nature 1997;390:45–51.

82 Kuro-o M: Klotho and aging. Biochim Biophys Acta 2009;1790:1049–1058.

83 Suzuki H, Amizuka N, Oda K, Li M, Yoshie H, Oshima H, Noda M, Maeda T: Histological evidence of the altered distribution of osteocytes and bone matrix synthesis in *klotho*-deficient mice. Arch Histol Cytol 2005;68:371–381.

84 Suzuki H, Amizuka N, Oda K, Noda M, Oshima H, Maeda T: Histological and elemental analysis of impaired bone mineralization in klotho-deficient mice. J Anat 2008;212:275–285.

85 Kawaguchi H, Manabe N, Miyaura C, Chikuda H, Nakamura K, Kuro-o M: independent impairment of osteoblast and osteoclast differentiation in klotho mouse exhibiting low turnover osteopenia. J Clin Invest 1999;104:229–237.

86 Kuro-o M: Klotho as a regulator of oxidative stress and senescence. Biol Chem 2008;389:233–241.

87 Suzuki H, Amizuka N, Oda K, Noda M, Oshima H, Maeda T: Involvement of the klotho protein in dentin formation and mineralization. Anat Rec 2008; 291:183–190.

88 Liu H, Ferguson MM, Castilho RM, Liu J, Cao L, Chen J, Malide D, Rovira Il, Schimel D, Kuo CJ, Gutkind JS, Hwang PM, Finkel T: Augmented Wnt signaling in a mammalian model of accelerated aging. Science 2007;317:803–806.

89 Verstraeten VL, Broers JL, Ramaekers FC, van Steensel MA: The nuclear envelope, a key structure in cellular integrity and gene expression. Curr Med Chem 2007;14:1231–1248.

90 Mounkes LC, Kozlov S, Hernandez L, Sullivan T, Stewart CL: A progeroid syndrome in mice is caused by defects in A-type lamins. Nature 2003;423:298–301.

91 Akter R, Rivas D, Geneau G, Drissi H, Duque G: Effect of lamin A/C knockdown on osteoblast differentiation and function. J Bone Miner Res 2009;24: 283–293.

92 Rivas D, Li W, Akter R, Henderson JE, Duque G: Accelerated features of age related bone loss in zmpste24 metalloproteinase deficient mice. J Gerontol A Biol Sci Med Sci 2009;64:1015–1024.

93 Le Gall JY, Ardaillou R: The biology of aging. Bull Acad Natl Med 2009;193:365–402.

94 Allori C, Sailon AM, Warren SM: Biological basis of bone formation, remodeling and repair. II. Extracellular matrix. Tissue Eng 2008;14:275–283.

95 Wang J, Lida K, Egusa H, Hokugo A, Jewett A, Nishimura I: Trabecular bone deterioration in col9a1+/– mice associated with enlarged osteoclasts adhered to collagen IX deficient bone. J. Bone Miner Res 2008;23:837–849.

96 Kato N, Kitahara K, Rittling SR, Nakashima K, Denhardt DT, Kurosawa H, Ezura Y, Noda M: Osteopontin deficiency enhances anabolic action of EP4 agonist at a sub-optimal dose in bone. J Endocrinol 2007;193:171–182.

97 Ikeda T, Nagai Y, Yamaguchi A, Yokose S, Yoshiki S: Age-related reduction in bone matrix protein mRNA expression in rat bone tissues: application of histomorphometry to in situ hybridization. Bone 1995;16:17–23.
98 Young MF, Bi Y, Ameye L, Chen XD: Biglycan knockout mice: new models for musculoskeletal diseases. Glycoconj J 2002;19:257–262.
99 Chen XD, Shi S, Xu T, Robey PG, Young MF: Age-related osteoporosis in biglycan-deficient mice is related to defects in bone marrow stromal cells. J Bone Miner Res 2002;17:331–340.
100 McBeath R, Pirone DM, Nelson CM, Bhadariraju K, Chen CS: Cell shape, cytoskeletal tension and RhoA regulate stem cell lineage commitment. Dev Cell 2004;6:483–495.
101 Windahl SH, Hollberg K, Vidal O, Gustafsson JA, Ohlsson C, Andersson G: Female estrogen receptor beta–/– mice are partially protected against age related trabecular bone loss. J Bone Miner Res 2001; 16:1388–1398.
102 Juttner KV, Perry MJ: High-dose estrogen-induced osteogenesis is decreased in aged RUNX2 (+/–) mice. Bone 2007;41:25–32.
103 Tsuji K, Komori T, Noda M: Aged mice require full transcription factor, Runx2/Cbfa1, gene dosage for cancellous bone regeneration after bone marrow ablation. J Bone Miner Res 2004;19:1481–1489.
104 Ducy P, Starbuck M, Priemel M, Shen J, Pinero G, Geoffroy V, Amling M, Karsenty G: A Cbfa-1 dependent genetic pathway controls bone formation beyond embryonic development. Genes Dev 1999;13:1025–1036.
105 Akune T, Ohba S, Kamekura S, Yamaguchi M, Chung UI, Kubota N, Terauchi Y, Harada Y, Azuma Y, Nakamura K, Kadowaki T, Kawaguchi H: PPARgamma insufficiency enhances osteogenesis through osteoblast formation from bone marrow progenitors. J Clin Invest 2004;113:846–855.
106 Wu XB, Li Y, Schneider A, Yu W, Rajendren G, Iqbal J, Yamamoto M, Alam M, Brunet LJ, Blair HC, Zaidi M, Abe E: Impaired osteoblastic differentiation, reduced bone formation, and severe osteoporosis in noggin-overexpressing mice. J Clin Invest 2005;112: 924–934.
107 Baylink j, Lau KH-W, Mohan S: The role of IGF system in the rise and fall in bone density with age. J Musculoskelet Neuronal Interact 2007;7:304–305.
108 Karsenty G, Elefteriou F: Neuronal regulation of bone remodeling; in: Primer on the Metabolic Bone Diseases and Disorders of Mineral Metabolism, ed 7. Washington, American Society for Bone and Mineral Research, 2008, pp 56–60.
109 Ducy P, Amling M, takeda S, Priemel M, Schilling AF, Beil T, Shen J, Vinson C, Rueger JM, Karsenty G: Leptin inhibits bone formation through a hypothalamic relay: a central control of bone mass. Cell 2000;100:197–207.
110 Elefteriou F, Ahn JD, Takeda S, Starbuck M, Yang X, Liu X, Kondo H, Richards WG, Bannon TW, Noda M, Clement K, Vaisse C, Karsenty G: Leptin regulation of bone resorption by the sympathetic nervous system and CART. Nature 2005;434:514–520.
111 Bouxsein ML, Devlin MJ, Glatt V, Dhillon H, Pierroz DD, Ferrari SL: Mice lacking beta-adrenergic receptors have increased bone mass but are not protected from deleterious skeletal effects of ovariectomy. Endocrinology 2009;150:144–152.
112 Idris I, Sophocleous A, Landao-Bassonga E, Canals M, Milligan G, Baker D, vant Hof RJ, Ralston SH: Cannabinoid receptor type 1 protects against age related osteoporosis by regulating osteoblast and adipocyte differentiation in marrow stromal cells. Cell Metab 2009;10:139–147.
113 Hamrick MW, Ding KH, Pennington C, Chao YJ, Wu YD, Howard B, Immel D, Borlongan C, McNeil PL, Bollag WG, Curl WW, Yu J, Isales CM: Age related loss of muscle mass and bone strength in mice is associated with a decline in physical activity and serum leptin. Bone 2006;39:845–853.
114 Hamrick MW, Pennington C, Newton D, Xie D, Isales C: Leptin deficiency produces contrasting phenotypes in bones of the limb and spine. Bone 2004;34:376–383.
115 Teitelbaum SL: Bone resorption by osteoclasts. Science 2000;289:1504–1508.
116 Owen M: Marrow stromal cells. J Cell Sci 1988;10: 63–76.
117 Lufkin EG, Wahner HW, OFallon WM, Hodgson SF, Kotowicz MA, Lane AW, Judd HL, Caplan RH, Riggs BL: treatment of postmenopausal osteoporosis with transdermal estrogen. Ann Intern Med 1992;117:1–9.
118 Chamberlain G, Fox J, Ashton B, Middleton J: Concise review: mesenchymal stem cells: their phenotype, differentiation capacity, immunological featuresand potential for homing. Stem Cells 2007;25: 2739–2749.
119 Bellantuono j, Aldahmash A, Kassem M: Aging of marrow stromal (skeletal) stem cells and their contribution to age-related bone loss. Biochim Biophys Acta 2009;1792:364–370.
120 Gimble JM, Zvonic S, Floyd ZE, Kassem M, Nuttall ME: Playing with bone and fat. J Cell Biochem 2006; 98:251–266.

121 Nakashima K, Zhou X, Kunkel G, Zhang Z, Deng JM, Behringer RR, Crombrugghe BD: The novel zinc-finger containing transcription factor osterix is required for osteoblast differentiation and bone formation. Cell 2002;108:17–29.
122 Lazar MA: PPAR gamma, 10 years later. Biochimie 2005;87:9–13.
123 Beresford JN, Bennett JH, Devlin C, Leboy PS, Owen ME: Evidence for an inverse relationship between the differentiation of adipocytic and osteogenic cells in rat marrow stromal cell cultures. J Cell Sci 1992;102:341–351.
124 Kassem M, Abdallah BM, Saeed H: Osteoblastic cells: differentiation and trans-differentiation. Arch Biochem Biophys 2008;473:183–187.
125 Kang Q, Song WX, Luo Q, et al: A comprehensive analysis of the dual roles of BMPs in regulating adipogenic and osteogenic differentiation of mesenchymal progenitor cells. Stem Cells Dev 2009;18: 545–559.
126 Gaur T, Lengner CJ, Hovhannisyan H, Bhat RA, Bodine PV, Komm BS, Javed A, van Wijnen AJ, Stein JL, Stein GS, Lian JB: Canonical WNT signaling promotes osteogenesis by directly stimulating Runx2 gene expression. J Biol Chem 2005;280:33132–33140.
127 Abdallah M, Jensen CH, Gutierrez G, Leslie RG, Jensen TG, Kassem M: Regulation of human skeletal stem cells differentiation by Dlk1/Pref-1. J Bone Miner Res 2004;19:841–852.
128 Rifas L: The role of noggin in human mesenchymal stem cell differentiation. J Cell Biochem 2007;100: 824–834.
129 Tang QQ, Otto TC, Lane MD: Commitment of C3H10T1/2 pluripotent stem cells to the adipocyte lineage. Proc Natl Acad Sci USA 2004;101:9607–9611.
130 Ding J, Nagai K, Woo JT: Insulin dependent adipogenesis in stromal ST2 cells derived from murine bone marrow. Biosci Biotechnol Biochem 2003;67: 314–321.
131 Nosjean J, Boutin JA: Natural ligands of PPAR-gamma: are prostaglandin J(2) derivatives really playing the part. Cell Signal 2002;14:573–583.
132 Rauner M, Sipos W, Pietschmann P: Age dependent Wnt gene expression in bone and during the course of osteoblast differentiation. Age (Dordr) 2008;30: 273–282.
133 Moerman J, Teng K, Lipschitz DA, Lecka-Czernik B: Aging activates adipogenic and suppresses osteogenic programs in mesenchymal marrow stroma/stem cells: the role of PPAR-gamma 2 transcription factor and TGF-beta/BMP signaling pathways. Aging Cell 2004;3:379–389.
134 Almeida M, Ambrogini E, Han L, Manolagas SC, Jilka RL: Increased lipid oxidation causes oxidative stress, increased peroxisome proliferator-activated receptor gamma expression, and diminished pro-osteogenic Wnt signaling in the skeleton. J Biol Chem 2009;284:27438–27448.
135 Shockley KR, Lazarenko OP, Czernik PJ, Rosen CJ, Churchill GA, Lecka-Czernik B: PPARgamma2 nuclear receptor controls multiple regulatory pathways of osteoblast differentiation from marrow mesenchymal stem cells. J Cell Biochem 2009;106: 232–246.
136 Okazaki R, Inoue D, Shibata M, Saika M, Kido S, Ooka H, Tomiyama H, Sakamoto Y, Matsumoto T: Estrogen promotes early osteoblast differentiation in mouse bone marrow stromal cell lines that express estrogen receptor alpha or beta. Endocrinology 2002;143:2349–2356.
137 Imam A, Iqbal J, Blair HC, Davies TF, Huang CL, Zallone A, Zaidi M: Role of pituitary-bone axis in skeletal pathophysiology. Curr Opin Endocrinol Diabetes Obes 2009;16:423–429.
138 Abdallah M, Haack-Sorensen M, Fink T, Kassem M: Inhibition of osteoblast differentiation of mesenchymal stem cells by sera obtained from aged females. Bone 2006;39:181–188.
139 Post S, Abdallah BM, Bentzon JF, Kassem M: Demonstration of the presence of independent pre-osteoblastic and pre-adipocytic cell populations in bone marrow derived mesenchymal stem cells. Bone 2008;43:32–39.

Farhan A. Syed, PhD
Abbott Labs
Worcester MA 01605 (USA)
Tel. +1 508 688 3532, Fax +1 508 754 6742, E-Mail farhan.syed@abbott.com

Author Index

Subject Index